The

ADDICTION

Inoculation

Also by Jessica Lahey

The Gift of Failure

The
ADDICTION
Inoculation

RAISING **HEALTHY KIDS** IN A CULTURE OF DEPENDENCE

Jessica Lahey

HARPER
An Imprint of HarperCollinsPublishers

The names and identifying details have been changed to protect the privacy of children described in this book.

This book is written as a source of information about substance use disorder and children. It is based on the research and observations of the author, who is not a medical doctor. The information contained in this book should by no means be considered a substitute for the advice of a qualified medical professional, who should always be consulted if you have or suspect you have a health problem.

The information in this book has been carefully researched, and all efforts have been made to ensure accuracy as of the date published. The author and the publisher expressly disclaim responsibility for any adverse effects arising from the use or application of the information contained in this book.

THE ADDICTION INOCULATION. Copyright © 2021 by Jessica Lahey. All rights reserved. Printed in the United States of America. No part of this book may be used or reproduced in any manner whatsoever without written permission except in the case of brief quotations embodied in critical articles and reviews. For information, address HarperCollins Publishers, 195 Broadway, New York, NY 10007.

HarperCollins books may be purchased for educational, business, or sales promotional use. For information, please email the Special Markets Department at SPsales@harpercollins.com.

FIRST EDITION

Designed by Nancy Singer

Library of Congress Cataloging-in-Publication Data has been applied for.

ISBN 978-0-06-288378-0

21 22 23 24 25   LSC   10 9 8 7 6 5 4 3 2 1

For my children and my students

CONTENTS

The

ADDICTION

Inoculation

CHAPTER 1

HI, MY NAME IS JESS, AND I'M AN ALCOHOLIC

Hi, my name is Jess, and I'm an alcoholic.

It has taken me such a long time to arrive at this sentence, to be able to put the word "alcoholic" in such close proximity to the word "I" and to face the reality that no matter how diligent I am, I can't control my drinking and must live a completely sober life. Eight years on in my recovery, I'm grateful to be here, beyond the shame, guilt, secrets, and lies. Now that my own relationship with addictive substances is well in hand, it's time for me to figure out how to prevent my children from having to travel the same path.

I come by my addictions honestly, as the branches of my family tree hang heavy with substance use disorder. Some of my relatives favored moonshine, others wine and pills, but the one constant was secrecy. No one ever talked about it. They stocked up on bulk packages of peppermint gum and hid their emergency nips in the rafters of the basement workroom while pretending everything was fine and dandy. If questioned, we were all great, thanks for asking, nothing to see here.

But there was plenty to see, even if we were not allowed to call it by its name. My family's native tongue was one of euphemisms, and I've never been a fan of the dialect. I was raised to understand that the proper term for passing out was "taking a nap," and voicing concerns about a relative's drinking or pill use was a punishable offense. In the midst of all this obfuscation and chaos, the only thing I hated more than alcohol was the lying. By the time I hit adolescence, I'd begun to understand the scope of my extended family's problem with addictive substances, and I was scared to death. Drugs and alcohol threatened my identity as a perfectionist, an overachiever, the goody-two-shoes eldest daughter, so I bolted to the abstinent end of the substance use spectrum and held on for dear life. I was sure of one thing in the way only teenagers can be certain: I would *never* grow up to be like them.

Except, I did.

Single-minded determination to thwart my genetic legacy did not come with operating instructions, so I made the rules up as I went along. It's no wonder I ended up at the bottom of a wine bottle in my forties. Even so, I was one of the lucky ones. I emerged relatively unscathed from my years of substance abuse. As a "not yet" or "high bottom" alcoholic, I did not have to lose my family, friends, or career in order to find my way out, and for that, I am immensely grateful. I married a man who shares my genetic predisposition for addiction, though he escaped that fate himself. We have two boys, and while we can't do anything to mend their genetics, we can promise them this: the language of shame, secrecy, and euphemism will have no place in our home.

I was a middle and high school teacher for twenty years, and spent the last five years of it teaching some of New England's most addicted children in an inpatient drug and alcohol rehabilitation center in Vermont. I was their writing teacher, charged with helping them find the words to pin their deepest, darkest monsters down on the page, to expose those ugly and manipulative beasts to the light and identify their parts. When the work was hard, I wrote alongside them, and together we learned how to describe the events that led all of us to that small rehab classroom, from our first use to our last hurrah and everything in between.

No one wants to grow up to be a drug addict or an alcoholic; that's simply where some of us end up, so desperate to escape the discomfort of being who we are that we pick up that first smoke or drink. I first drank because I was anxious to impress a girl I admired, but I wish I'd known then that she first drank because she disliked herself so much that she wanted to disappear.

My friend Celeste (not her real name) and I were alone in her house one weekend night when she came up with a plan. We'd take one sip out of every bottle in her parents' liquor closet. It would be enough to get drunk, but not enough to get caught. I went along with her plan but most of my sips were pretend. Hers, however, were in earnest, and the night went about as well as one might imagine. She got drunk, I got tipsy. She threw up; I held her hair back, cleaned her up, and put her to bed.

Celeste and I went our separate ways after middle school, but in an internet-fueled fit of nostalgia, I searched for her. I'd expected to find her Facebook page, images of her face on some law firm website, or a mention of her marriage in a local paper. What I found instead was her obituary.

It offered no information on her cause of death, only that she'd died at thirty-five, leaving her parents, siblings, and an ex-husband,

and in lieu of flowers they asked for donations to the American Liver Foundation. Facebook led me to her sister, Anne (also a pseudonym), who told me that Celeste died of liver failure after a multi-year, *Leaving Las Vegas*–style drinking binge. After a couple of failed suicide attempts, Anne said, Celeste chose alcohol as her surest path to death and drank as much as she could in order to get there as quickly as possible.

The impulsive, funny girl I'd so wanted to impress in middle school grew into an explosive, erratic young adult whose pattern of broken relationships and uncontrolled mood swings culminated in a diagnosis of borderline personality disorder (BPD) at eighteen. Unfortunately, by then it was too late to save her. Her sister, who still does not fully understand Celeste's disease or suicide, did her best to explain it to me.

> If you look up the behaviors [or BPD] online, she was a classic case. Taken one at a time the signs made her look like a typical teen but when added all together, [her diagnosis] became clear. The problem was nobody put all the pieces together until she was over 18 and nobody could force her to do any treatment. She used alcohol to self-medicate.

It's no surprise that Celeste turned to alcohol to ease her symptoms. Both BPD and substance use disorder are characterized by emotional dysregulation and impulsivity, as well as feelings of emptiness, instability, and the inability to control moods. Many people with BPD turn to alcohol to medicate these symptoms, and because Celeste's BPD went untreated even after she was officially diagnosed, alcohol provided relief, at least in the short term.

Celeste died just as my daily wine intake was ramping up. We were two alcoholics traveling two very different paths, but our reasons for drinking were the same: to self-medicate our emotional and psychological discomfort. Mine stemmed from anxiety disorder and compulsive perfectionism, Celeste's from her untreated borderline personality disorder, both predictable, well-known causes of alcohol dependence and abuse.

Anxiety and substance use disorder are familiar bedfellows, especially in women. Women are twice as likely to have anxiety disorder as men, and women with anxiety disorders are more likely to use alcohol to self-medicate their symptoms. In fact, women with anxiety disorders are more likely to abuse alcohol than to drink normally, and progression from alcohol use to dependence happens faster in these women. I would have done just about anything to escape my anxiety, but all I had until I was diagnosed and treated in my twenties was alcohol. The only thing that kept me off Celeste's path was my abject fear of becoming an alcoholic, but over the long term, even that could not save me.

Alcohol is such a trickster; it works so well in the short term to quiet the symptoms of both anxiety and BPD, but over the long term, it exacerbates both conditions. Alcohol compounded my anxiety and worked against the medication I took to manage it. As the consequences of my drinking began to mount up, they added to my anxiety, which made me want to drink more. Celeste found no cure at the bottom of her bottle, either. Her BPD continued to eat away at her happiness and well-being, and once her marriage failed, she no longer drank to feel better, she drank to obliterate herself.

I wish I'd known on that evening in her parents' house what lay ahead for both of us. I wish I'd been strong enough to tell her I was not comfortable being a part of that evening's drinking experiment. I wish I'd told her I loved her, that she had the power to make other people happy simply by being herself. I wish I'd known she was beginning a descent into a madness that would become full-blown mental illness. More than anything else, I wish she'd been diagnosed in time to replace her attempts at self-medication with real treatment that could have saved her life.

We attended the same high school, but as we were no longer friends, I didn't know how much she drank. I was a teenaged teetotaler. I pretended to drink from time to time, even pretended to be drunk once or twice, just to fit in, but mostly, I was the perpetual designated driver, responsible for making sure everyone else got home safely. I was a dependable source of mints and gum and the keeper of the car keys.

In my second year of college, I went all in on sobriety and righteous self-satisfaction and trained to become a peer drug and alcohol counselor. I was that annoying holier-than-thou twerp my college's health services sent in to reeducate busted frat boys. Picture a twenty-year-old Jess, standing at the front of a dank, beer-soaked University of Massachusetts fraternity common room on a sunny Sunday morning. Thirty or forty grumpy young men take their seats on the damp and sagging furniture, and snicker as Jess mounts her colorful, educational flip charts and illustrative diagrams on a portable aluminum easel. For the next hour, she uses every trick in her very short book to sway the hearts and minds of these floridly pre-alcoholic young men while they attempt to keep themselves awake with coffee, fidgeting, and mental plans for next week's kegger, the one Jess will most definitely not be invited to.

I learned everything about the science of addiction but nothing about my own relationship to it, let alone how to defend myself against it. I could explain the biochemistry of sober to shitfaced with my pointer and a diagram: 20 percent of the ethanol in their cheap kegged beer or vile house punch is absorbed in the stomach before moving on to the intestine, where the other 80 percent escapes into the blood. Once it hits the liver, an enzyme called alcohol dehydrogenase breaks the ethanol down into acetaldehyde, and another enzyme called aldehyde dehydrogenase breaks the acetaldehyde into acetate, which leaves the body in the breath that condenses on the frat windows and the pee they use to write their names in the snow.

I graduated from college with my sobriety intact, but I'd started to fall in love with the romance of drinking, and romance turned out to be the thin edge of my addiction wedge. There's a scene in the movie *Raiders of the Lost Ark* where Marion, the headstrong, willowy barkeep, wins a drinking contest against a massive Nepalese man. After many, many shots, he passes out, but somehow Marion is completely sober as the crowd cheers and she gathers up her winnings. I loved that scene. Marion is so badass she can drink a huge man under the table, then go on to put her long-lost love Indiana Jones in his place with barbed banter. In reality, of course, that could never happen. If the big guy is unconscious, she'd be long dead, but who cares? I loved the idea of being able to outdrink the most hardened locals in a far-flung bar. Later, when I read the *Outlander* books, the hook was set. Jamie and Claire savored their soothing drams of whiskey by eighteenth-century firelight and drank down their stream-chilled stone jugs of cider after haying in the August heat. I'd always thought that hard cider tasted like carbonated, rotten hot dogs, but the taste wasn't really the point. I wanted to drink deep from the romance, the taste of Jamie and Claire's life in the Scottish Highlands and the mountains of North Carolina. It was simple, and beautiful and good, and until I could have their life, their drink would do.

The romance began to wear off in my thirties, and I was thirty-two when I realized I might have a problem, or at least the very beginnings of one. My husband and I lived in Cambridge, Massachusetts, where I worked as a nanny in exchange for an attic apartment while he completed his medical fellowship. My older son was five, the little girl I nannied was a toddler, and my second son was an infant. I never drank during the day, but I really looked forward to the moment I handed the child over to her mother and I could open that evening bottle of wine. I drank while cooking elaborate meals for my family, and I drank while we ate those delicious meals. At some point during those years, one bottle of wine split between the two of us was no longer just the right amount, and that made me nervous. Not nervous enough to throttle back my drinking, but nervous enough to ensure I always had a backup bottle. We did not really have the money for multiple bottles of wine, but I rationalized the purchase somehow. I cooked a lot of things that required a splash of wine and provided an extra glass or two or three for me, and for a while, that solved the issue of supply and demand.

My emerging problem, however, was disposal. I had to be careful. Very, very careful.

One afternoon, I was leaning up against the brick wall of my son's elementary school at kindergarten pickup, and another parent, a woman I barely knew, approached. She clearly had something to say and was having trouble getting it out. She leaned toward me, took a breath, then opened her mouth and closed it again. Opened her mouth, and closed it once more. I braced myself because she was hemming and hawing so much I thought she was going to tell me my son had been mean to her kid.

"Can I ask you something?" she asked.

"Sure," I said, relieved, because my son being mean would be a statement, not a question.

"If I really look forward to a glass or two of wine at the end of every day, do you think that's a problem? I mean, does that make me an alcoholic?"

I was a virtual stranger to this woman and she was asking me to make a clinical judgment about her substance use. I figured my role in this exchange was to offer her one of two options:

OPTION ONE: "No! Of course not! One or two glasses of wine at night is supposed to be good for you, right? I mean, especially if it's red wine. Resveratrol. It fights cancer or something. Besides, the French drink wine with lunch and dinner, so no. Totally not. You are completely normal. In fact, you deserve a glass or two at the end of the day."

OPTION TWO: "I guess it could. Maybe the fact that you are worried about it is a good enough reason to talk to someone."

I don't know which answer she wanted, but the first option was the only one I was emotionally prepared to face myself, so we spent the next half hour talking about resveratrol and the romance of Paris.

Neither of us was ready for option two. Not yet.

That night, after my husband and kids were asleep, I googled "test" and "am I an alcoholic?" because if there's one thing overachieving perfectionists do well, it's take tests.

1. Have you ever decided to stop drinking for a week or so, but it only lasted for a couple of days?

    Well, sure, but I stayed stone-cold sober both times I was pregnant. Didn't drink a drop.

2. Do you wish people would mind their own business about your drinking—stop telling you what to do?

   Yes, but I don't like anyone telling me what to do, about anything.

3. Have you ever switched from one kind of drink to another in the hope that this would keep you from getting drunk?

   No, but I have switched from beer or wine to liquor in order to get drunk faster.

4. Have you had to have an eye-opener upon waking during the past year?

   Ew, no. And not that I'm keeping track, but that's two "no" answers so far. Phew.

5. Do you envy people who can drink without getting into trouble?

   Do those people exist? Huh. It hadn't occurred to me before, but sure, of course, it would be nice to let go every once in a while, with no consequences.

6. Have you had problems connected with drinking during the past year?

   No, not really. No. Well . . . no.

7. Has your drinking caused trouble at home?

   It's caused conversations, but not "trouble," really. Come on, test. Define your terms.

8. Do you ever try to get "extra" drinks at a party because you do not get enough?

   Okay, fine, yes, especially when I don't have a chance to drink before the party. But the hardest part of the party is walking in the door, so why wouldn't you start ahead of time?

9. Do you tell yourself you can stop drinking anytime you want to, even though you keep getting drunk when you don't mean to?

    Yes, but I did. I did stop for both pregnancies, so I can stop anytime I want to. I'm just not sure I want to anymore.

10. Have you missed days of work or school because of drinking?

    No.

I scrolled down to the bottom of the quiz to find out how many "yes" answers merit the label "alcoholic." "Did you answer YES four or more times? If so, you are probably in trouble with alcohol." I went back up the list to do a recount of questions one through ten before moving on to the final two, using a system of half points where I felt partial credit was warranted. I was up to a 4.5, or maybe 5, which is hardly definitive of anything.

11. Do you have blackouts?

    Yes.

12. Have you ever felt that your life would be better if you did not drink?

    Yes.

What I should have done the moment I finished that quiz was google "twelve-step meetings near me," seek out a therapist, tell my husband, tell my doctor, do just about anything other than what I actually did, which was to double down on my right to drink.

I was overcome with what another alcoholic, author Stephen King, calls "frightened determination." When he looked down on the plastic bin of empty sixteen-ounce cans of Miller Light in his garage and realized, "Holy shit, I'm an alcoholic," he doubled down, too.

> My reaction to this [realization] wasn't denial or disagreement; it was what I'd call frightened determination. You have to be careful, then, I clearly remember thinking. Because if you fuck up—
>
> If I fucked up, rolled my car over on a back road some night or blew an interview on live TV, someone would tell me I ought to get control of my drinking, and telling an alcoholic to control his drinking is like telling a guy suffering the world's most cataclysmic case of diarrhea to control his shitting.

All I had to do was be careful, and I could do that. I was the most careful person I knew, magna cum careful, and for about a decade, careful worked. My drinking increased so slowly I hardly noticed it was happening. Eventually, though, those one or two glasses of wine a night turned into two or three and by the time I hit forty-four, I was drinking at least a bottle of wine a day, sometimes augmented by whatever else we had in the house. The whiskey I bought "as a present" for my husband, Tim, was fair game, as was the vodka I used to make homemade vanilla with crushed vanilla beans. I drank in the afternoons, after I got home from teaching middle school but before Tim got home from work. In service to my careful planning, I developed all kinds of rituals around hiding and protecting my right to drink.

I used to purchase a bottle of red wine at our local market on my way home from school, swearing I'd share it with my husband. Just one glass today, I'd promise myself. I'd pour that one glass in a very large goblet that held at least a half a bottle, then empty the rest into a quart-sized mason jar. I'd then stash that mason jar in the freezer. Why a mason jar? Why the freezer? In theory, it was because I'd probably make a beef stew at some point, and if I needed the half bottle of red wine, it would be there.

But I hardly ever made beef stew.

Even if I'd wanted to make beef stew, there would not have been any red wine to make it with, because about an hour after I had that first glass, I'd pull that jar out of the freezer, stir up the frozen slurry, and drink it, ice chunks and all, like a red wine Slurpee. Sometimes I microwaved it to speed the thaw. It tasted weird, but whatever. At that point, I wasn't ever in it for the taste. By the time Tim got home from work, I was already a bottle of wine in, so if he'd done as I'd asked, and stopped by the market for a bottle of wine to drink with dinner, I'd be full-on drunk by bedtime. I often needed help getting to sleep, and the wine worked great.

Unfortunately, it was also great at waking me up at around three every morning. Alcohol feels like a sleep aid because it's a sedative hypnotic. It puts you to sleep at first, but it blocks REM sleep, the most restorative sleep phase, and causes late-night (or in my case, early-morning) wakefulness. I'd fall into a nice, relaxing boozy slumber at ten, but at three, I'd emerge from a sound sleep and begin the laborious and stressful process of piecing together the scattered fragments of the night before. By the end of my drinking, I was blacking out often and had to work hard to hide my memory lapses. What did we talk about? Is there anything I have to remember for tomorrow? Did I do anything stupid? Did I call anyone? Can I blame it on fatigue, or stress, maybe convince my family I was coming down with something? If stress over the blacking out snowballed into a full-on anxiety attack, I was done with sleep for the rest of the night.

Ah, yes. My anxiety disorder. It had been fairly well controlled with medication since my mid-twenties, when my physician and I finally found a drug that kept the anxiety attacks at bay without side effects. My anxiety medication of choice is an SNRI, or serotonin-norepinephrine reuptake inhibitor. It isn't as dangerous in combination with alcohol as, say, benzodiazepines, but it definitely falls under "not at all physician-approved and generally suboptimal."

I made the same vow almost every night for two years: Never again. I'm done. That's it. No wine tomorrow.

I broke that vow nearly every day over those two years. Each new limit—one bottle, never before sunset, never at lunch, never in front of my kids—fell victim to my growing thirst and waning control. I drank more when I was anxious, less when I was on an even keel, but I always drank. My daily goal was to get enough alcohol on board without stumbling over the line between happily buzzed and conspicuously drunk. My aim for that line was pretty good. I was a chatty, happy drunk, and unless I burned dinner or wandered over to visit the neighbors with an apron full of tomatoes from the garden and forgot to go home, I looked fairly normal on the outside. Tim kept a close eye on me given our shared family history, but I was determined to elude detection.

Tim is a physician and an educator who has received many awards for his empathetic and thoughtful teaching and mentoring. It is his job to put people at ease so they will share the clues that reveal the truth of their health and well-being. He sees people clearly, even when they don't want to be seen. Keeping the full truth of my drinking from him required a great deal of planning and attention. I was so good at hiding my buzz that when I started telling my friends I'd gone into recovery, I found myself in the odd position of having to convince my closest friends that no, really, I do have a problem with alcohol.

An extremely high-functioning alcoholic, I had two full-time jobs, one as a teacher with a nightmare schedule of seven different preps—Latin 6, 7, and 8, English 7 and 8, and writing 7 and 8—and a second writing about education and child welfare for the *Atlantic* and the *New York Times*. I always had papers to grade, an article to write, edits to check, and classes to prepare. In order to make this schedule work, I had to abide by a lot of rules. I never drove drunk, so I had to schedule my drinking very carefully. I never graded papers or submitted articles when I'd been drinking, which meant that as my drinking ramped up, I had to squeeze all my productivity into fewer and fewer hours of the day. I worked early, late, and in every spare moment between classes. I had to be in charge of our recycling because I could not risk anyone else seeing what I'd tucked under the laundry detergent jugs and cans of black beans. I had to keep track of my hiding places for cans of emergency beer, bottles of wine, not to mention the secret bottle of scotch I used to top up the public bottle of scotch we kept on the top shelf of our pantry.

By June 7, 2013, I was exhausted.

After years of dreading blackouts, I am grateful I have no memory of what happened the night of June 6, 2013, when I hit my personal bottom. That night, we celebrated my mother's birthday with family and friends who had gathered from far and wide on a gorgeous night in Rockport, Massachusetts. My oldest friend, someone I love as a sister and had not seen in years, had flown in to celebrate with us. As we sat down to an elaborate dinner, my father poured champagne for everyone to toast, but my friend demurred. One of my last clear memories from that night was her explanation: "All of the bad decisions I've made in my life were made with alcohol, and I decided drinking just wasn't worth it anymore." After that last moment of clarity, things got blurry really fast.

The pictures of that evening are gorgeous. The kids playing in the yard, friends gathered on the porch and walking on the rocks by the ocean, but the time stamps end around nine, when I got too drunk to take any more pictures. Apparently, Tim took me up to bed around eleven. Early the next morning, while Tim was out on a run, my father knocked on the bedroom door. He came in, sat at the end of my bed, looked me in the eye, and said, "I know what an alcoholic looks like, and you are an alcoholic."

All I had left in me was relief and assent. "I know," I said, then waited for him to leave the room so I could run into the bathroom and throw up.

That night, I attended my first twelve-step meeting. I was so panicked by the prospect of running into someone I knew from our small New Hampshire town that I drove forty minutes away to a large meeting in Vermont. I sat in the parking lot, watching the straggling smokers stub their cigarettes out, then fell in behind them as they headed inside the church and down into the basement. The room was full, and the only seat left was up front, with four other women. I'd felt pretty stable walking in, but the moment I sat down, I started crying. Literally, my butt hit the chair, and the tears started to flow. I cried through the introductions, the twelve traditions, and the first speaker. I cried through the secretary's report, the announcements, and the second speaker. The woman across from me did her best not to stare as she passed me tissue after tissue from some endless supply at the bottom of her gigantic purse.

At the very end of the meeting, a man named D. volunteered to hand out the chips.

"It is a tradition in this group to hand out chips for varying lengths of sobriety, and we always start with the most important one. Would anyone like to take a twenty-four-hour chip, the promise of a first day sober?"

I don't remember making the decision to get up and accept that silver plastic chip, but I did. I can't imagine I said anything, but I'm sure a lot of snot was involved if I did. That meeting, selected for its significant distance from my house, became my home group, and despite the miles and time involved, I attended nearly every Sunday for the first five years of my sobriety.

Two days after that first meeting, I wrote an essay that was later published at the *Huffington Post* about coming out to my family as an alcoholic and my willingness to label myself as such. That essay concludes,

> My son introduces me to his friends as many things—mother, wife, writer—and I'm incredibly proud of those labels. Proud enough that I refuse to allow this newest label to obliterate everything else I've worked so hard to become. I've finally done the math and figured out that the only way I get to keep those other identities is to admit the word "alcoholic" to my list.
>
> Because when my son is my age, I want him to be proud of me, particularly if our mutual inheritance grabs hold and threatens to drag him down. As his mother—particularly his alcoholic mother—the most important gift I can give him is the power of my example to guide him if he ever stumbles upon the treacherous terrain of our family's well-worn slippery slope.

That slippery slope is still there, as treacherous as ever, and I keep a constant eye on it. In the meantime, sobriety feels like my new normal, and that's been a huge relief. I'm allowed to be forgetful, daft, or uncoordinated, which is infinitely less humiliating than drunk. I was hit by a snowplow in early 2019 as I was driving away from a friend's house, and sitting behind the wheel of my wrecked car, shaking from the sudden flood of adrenaline, I was once again grateful for sobriety and the clarity of simple cause and effect. Before I got sober, friends and family would have wondered, did anyone breathalyze Jess? Might her response time have been impaired? Best of all, now that I don't have to stand guard over the buried secrets at the bottom of our recycling bins, I've handed trash duty over to Tim. Seems like a fair trade to me. I gave up booze, so he gets to do the recycling and he doesn't have to divorce me.

Because he would have. He told me so, sometime after our first son was born. He, too, knew what an alcoholic looked like, and he said he would not raise children amid that kind of sadness, shame, and worry. At the time, his statement was posed as a hypothetical, but it wasn't, not really. We've both known for a long time that we carry the legacy of our families' addiction in our bodies, that it's right there, waiting for one of us to fall asleep on the watch. Tim has been very careful, and has never let his guard down, but I did. I swung the front door wide open and invited it into our home.

I got so very, very lucky. My boys were young enough that they claim to have no memory of my drinking. My older son slept through my dry heaving in the bathroom right outside of his bedroom door. My younger son did not catch me the day he wandered into the garage just as I was stealing a nip from my hiding place. They are safe from my alcoholic past, so my worry has shifted to their futures. Now that I'm sober, with clarity to spare, all I can see is their inherited risk, hanging overhead like a sword of Damocles.

I am a teacher and researcher at heart, so I read everything I could find on the subject of substance abuse and child development—textbooks, how-tos, memoirs, and academic papers. As I read and took notes in the margins, I peeked over the top of the books, watching my kids for signs such as aggression, excessive sensation-seeking, and a craving for risk. I scored my kids like I scored that "Am I an Alcoholic?" quiz. Add one point for a love of roller coasters; take one away when they both answer no, roller coasters aren't their thing.

I scored their friends, watching for signs of potential addictive behavior, because I'd heard substance use among peers is a big risk factor for substance use. My older son, Ben, had one friend who was lovely but disconcertingly attracted to risk, constantly pulling boneheaded stunts like riding his bike off cliffs and constructing elaborate luge tracks through the woods behind his house. Those steeply banked chutes and hairpin turns gave me nightmares all winter long.

When my younger son fell in love with crystal healing, Pink Floyd, and psychedelic posters, Ben took me aside and said, "You know who else loves crystals and Pink Floyd and psychedelic posters? Potheads." The observation was unnecessary, as I'd spent the past year sniffing beneath the wafting incense smoke for the skunky stink of pot. Can a love of crystals and *Dark Side of the Moon* cause substance use disorder?

Between lessons in my drug rehab classroom, I conducted a fact-finding mission. I adored my students, even with all their challenges and button-pushing, and I looked to them for some kind of formula for predicting childhood substance abuse, maybe an algorithm I could use at home, one that, given enough variables, would spit out a probability figure. My students taught me so much about educating people whose emotional needs often eclipse all else, people who carry adult-sized burdens before their age of majority. They also explained how the dark web functions in the drug trade and which New England towns supply the best drugs. According to a note I intercepted as it was being passed across the classroom aisle, "Montpelier has lots of weed, coke, Molly, LSD, shrooms, but no dope. Barre, though . . . so much dope, crack, pills, etc." I keep that note in a folder with the lesson plan for that day and a grammar unit on adverbs.

When we lived in New Hampshire, and I watched the states around us decriminalize marijuana, I was politically and socially supportive while parentally freaking out and adding points to my kids' risk scores. In the years since states started legalizing, marijuana use among kids 12–17 and 18–25 has continued to climb. Sure, there's a predictable bump for novelty that might wear off in a few years, but now is when my kids' brains are more vulnerable to the damage pot does to memory and learning. Last year, when we moved from New Hampshire to Vermont, where recreational marijuana use is legal, I added another point to their scores.

Given all of these genetic and environmental risk factors for substance abuse, my best approach as a parent is to cop to our family history without shame or guilt, talk openly about my own substance abuse, and do my best to counter whatever other risk factors arise with as many protections as I can muster. We talk. We support. We offer their friends a comfortable and accessible place to hang out, and most of all, we listen.

After all my research, interviewing, attending substance abuse and mental health conferences, and teaching hundreds of addicted kids, here's what I have learned about preventing substance abuse in childhood and beyond:

HUMANS HAVE USED MIND-ALTERING SUBSTANCES FOR THOU-SANDS OF YEARS TO EASE OUR ANGST. If we can help kids manage their emotions and moods without having to resort to self-medication, we can increase their chances of making it to adulthood substance-free.

CHILDREN'S BRAINS ARE CELLULARLY, COGNITIVELY, AND FUNC-TIONALLY DIFFERENT FROM ADULT BRAINS. They respond differently to drugs and alcohol, and addictive substances do more damage in a child or adolescent brain that is still developing than in an adult brain.

THE YOUNGER KIDS ARE WHEN THEY START USING DRUGS AND ALCOHOL, THE MORE DAMAGE THEY DO TO THOSE BRAINS. Drugs and alcohol do both short- and long-term damage to the areas of the brain responsible for learning and memory.

THE YOUNGER KIDS ARE WHEN THEY START USING DRUGS AND AL-COHOL, THE MORE LIKELY THEY ARE TO DEVELOP SUBSTANCE USE DISORDER AS ADULTS. Adolescents who start drinking at twelve or younger are nearly four times as likely to develop substance use disorder as people who start drinking after eighteen.

ADOLESCENTS ARE BIOLOGICALLY WIRED TO SEEK OUT NOVELTY AND RISK. Effective substance abuse prevention tools take these neurological and cognitive realities into account and capitalize on them.

EFFECTIVE SUBSTANCE ABUSE PREVENTION REDUCES RISK FACTORS WHILE AMPLIFYING PROTECTIVE FACTORS. The best risk-reduction measures and protections are applied early and are tailored to the whole child, taking their age, gender, developmental stage, ethnicity, socioeconomic status, family, and community into account.

NO DEMOGRAPHIC IS SAFE FROM SUBSTANCE USE DISORDER. Over the years, the media and popular culture create stereotypes about what type of people are addicted to certain drugs, but the reality is that substance dependence doesn't care about media stereotypes. Addiction happens in every community, at every socioeconomic level, and every demographic.

GENETICS MATTER, BUT THEY ARE NOT DESTINY. Genetics are about 60 percent of the picture, and the rest comes down to environmental factors. While we can't do much about genetics, we do have the power to prevent, intervene, and treat kids for the environmental risks.

PREVENTION WORKS. Current data shows a continuing decline in substance use, abuse, and dependence among children and adolescents, and those declines are correlated with the rise of evidence-based prevention programs.

Before I got sober, I read addiction memoirs to give me hope and show me a way forward. Now that I'm sober, I continue to read them because they remind me of all I stand to lose if I start drinking again. But the one book I wanted most was the one I could not find, a memoir of substance use disorder and long-term recovery paired with research-backed parenting and teaching advice.

Once I'd faced my own addictions, I was desperate to learn more about my children's risk and understand the steps I could take to inoculate them against substance use disorder. I am so very, very lucky; I got sober before my life imploded, and I'd do just about anything to protect my children from the pain, anxiety, and sadness that come with this disorder. All kids are at some level of risk for substance abuse, but research shows that we can not only predict which kids are at higher risk due to biological or environmental factors, we can also prevent those kids from taking their first drink or drug.

So let's get to work.

CHAPTER 2

A LONG, STRANGE TRIP
Drugs, Alcohol, and Us

The students I taught over my five years at the inpatient drug and alcohol rehabilitation center for adolescents in Vermont would stay from a few weeks to a few months while they learned how to live their lives as sober people. The first days were always rough, as the numbness faded away and they had to face, then experience, the uncomfortable emotions and trauma they'd been working so hard to forget. I supported that work by making my classroom a safe place to remember. We hardly ever confronted their demons head-on; rather, we came at them tangentially, through metaphor and stories.

One of their favorite books was Jarrett J. Krosoczka's graphic memoir, *Hey, Kiddo: How I Lost My Mother, Found My Father, and Dealt with Family Addiction*, because they identified with the author's longing for love, stability, and a coherent history. The book recounts Krosoczka's childhood through words and pictures, first with a heroin-abusing mother, and later in the custody of his grand-parents, searching for a father he never knew.

I asked my students to bookmark moments in the text or illus-trations that resonated. Some chose the day Krosoczka goes to live with his grandparents; others selected the day he sees his father for the first time and realizes, "He was shorter than I imagined him to be." Then the kids created their own "memoir of a moment," combining text and pictures to describe a pivotal story from their own lives. They wrote what they could, drew when the words came hard, and the back-and-forth between these two expressive forms allowed difficult moments in their history to materialize on the page.

I watched Mark, a seventeen-year-old boy who had arrived at the rehab just a few days before, stare helplessly at the boxes he'd drawn on the page that would eventually contain the words and images of his story. His fingers traced the black lines, empty boxes floating in the white space of the page, and he quickly became overwhelmed, desperate for a starting place. Chaotic lives beget chaotic memories, not easily confined to neatly drawn lines and square angles.

I crouched down beside him and asked him to tell me a story, a short one, just to get started.

"I can't. There's too much, and it's all jumbled up," he said.

I opened *Hey, Kiddo* to the pages where Krosoczka reveals to his friend Pat that his mother is addicted to heroin. These facing pages tell a complete story in the author's life, I said. Tell one story, and you'll discover another, I promised.

So goes any attempt to chronicle human history. Truth, or what passes for it, is elusive, subject to the honesty and will of the teller. The story of how any one child moves from first use to substance abuse tends to follow a common trajectory, but the unique devils live in the details of nature and nurture. The story of our country's relationship to alcohol and drugs is similarly elusive, as it is subject to the bias, perspective, and agenda of the people who get to write it down. In this chapter, I offer my version of these histories.

The First Day

A few years ago, I watched Chris Herren speak to a couple of hundred high school students about his experience as a former NBA basketball player and recovering heroin addict. We were in a school gymnasium, an acoustically challenging space where every whisper and rustle and foot tapping on the bleachers reverberates as if it's amplified. As soon as Herren started speaking, however, the room fell silent. The kids were rapt. Some were stunned into silence, a few were weeping or trying their best not to cry, and many were shifting uncomfortably in their seats, but all were listening. Chris doesn't earn their attention with horror stories about the day he hit bottom (although his is a doozy). He reaches them because he goes deeper, recounting the story of his first drink, and the reasons behind it.

"We focus on the worst day and forget the first day," he tells them, because Chris doesn't care so much about the horrific places addiction takes kids in the end. He cares about the angst that gets them started. The stories we tend to tell about the horrors of substance abuse highlight the filth and desperation of an addict's final days but hardly ever detail the backstory, and it's the backstory that can help us prevent the first drink or drug from happening altogether.

In 1975, Denise Kandel and Richard Faust set out to investigate the natural history of substance use and abuse and the "gateway hypothesis," the commonly held belief that marijuana is the stepping-stone to heroin. While some details of their gateway hypothesis research have been criticized or debunked by more recent studies, they discovered some important patterns that can help us understand how kids progress in their substance use. Kandel and Faust surveyed high school students to discover how substance use begins, what substances they use and in what order, and how use evolves into abuse among adolescents. In their first paper about the survey, Kandel and Faust proposed an "initiation sequence" of gateway or stepping-stone drugs: 1) beer or wine; 2) cigarettes or hard liquor; 3) marijuana; and 4) other illicit drugs. Of the kids who go on to try marijuana, 98 percent started with beer or wine, and progression from alcohol to marijuana can be predicted based on the intensity and type of alcohol consumption. If, for example, a kid starts with beer then moves on to hard alcohol, they are much more likely to move ahead in the progression than a kid who sticks with beer. An important caveat is that just because a kid uses one substance, it does not necessarily mean they are going to progress to other substances.

Their findings reveal that when kids use nicotine and alcohol they are much more likely to progress to illicit drugs, and conversely, "Adolescents are very unlikely to experiment with marijuana if they have not experimented previously with an alcoholic beverage or with cigarettes; very few try illicit drugs other than marijuana without prior use of marijuana." More recent studies, however, take exception with Kandel and Faust's conclusions and describe a more nuanced version of their so-called gateway: while alcohol and nicotine use in adolescence may predict whether teens will move on to illicit drugs in the short term, predictions do not hold over the long term. "Over a relatively longer period of time (from adolescence to adulthood), early use of marijuana and other illegal drugs rather than tobacco or alcohol greatly increases the likelihood of using cocaine and other illegal drugs." In other words, kids who use nicotine and alcohol in adolescence are more likely to use harder drugs during adolescence, but if we want to know who will be most likely to progress to substance use in adulthood, we should be keeping an eye on the kids who use marijuana and other drugs.

When crack exploded on the drug scene in the 1980s, and rumors began to circulate in the media that crack was so alluring and addictive it had "broken" the gateway hypothesis and naive users were going straight into crack use without prior substance use experience, Kandel and her research partner Kazuo Yamaguchi surveyed nearly 8,000 students in grades seven through twelve in New York State public and private schools and concluded, no, crack has not broken the rules of substance use progression. No matter what the hyperbolic media broadcasts on the evening news, crack use follows the same old sequence: alcohol and/or cigarettes first, then marijuana, then, if it's going to happen at all, other illicit drugs, including crack.

That said, variations on this basic progression do exist, and different gender, cultural, and socioeconomic groups may exhibit slightly different patterns of use. For example, one study of Black and White boys recruited from Pittsburgh public schools found that both White and Black males are likely to begin their substance use with nicotine and/or tobacco, and if they progress, they move on to marijuana and hard drugs. Black males, however, are more likely to halt their progression at marijuana, whereas White males are more likely to escalate their use to hard drugs.

Another study found that while marijuana use usually follows alcohol use in the initiation sequence, this order is more likely to be reversed among Black young adults. Black adolescents and young adults, in particular, were three times as likely as White adolescents and young women to use marijuana before they use alcohol.

The takeaway from all of this: if we can keep kids away from cigarettes, vaping, beer, and marijuana, we are more likely to keep them off the harder stuff. This information can be helpful not only for parents who are hoping to keep an eye on their own children's risk for use and abuse, but it is extremely useful in designing prevention programs for schools and communities. If we know which substances act as gateway drugs, we can focus prevention efforts and keep kids from going down the initiation sequence altogether.

Effective prevention requires us to understand why a kid picks up that first mind-altering chemical, be it nicotine, alcohol, or cannabis, and address that cause, head-on. If a child drinks her first beer to escape memories of sexual abuse or to quiet the voices of self-hatred in her head, no amount of evidence about the harm booze can cause in her body will stop her. In Chris Herren's words, "If we can understand the beginning, we can help change the ending."

Humanity's first substance use, like that of most American teenagers, began with beer. In fact, there's mounting evidence that it was our desire for beer, not our need for bread, that helped us to evolve from small groups of nomadic hunter-gatherers into organized, civilized societies. Anthropologist Brian Hayden proposes it was the feasts, and the beer early humans served at those feasts, that allowed us to organize and thrive as a species. If we understand how beer eased the way for developing humanity, then we can better understand why our own developing humans use drugs and alcohol to ease adolescent angst.

Between 10,200 and 12,500 years ago, hunter-gatherers established settlements in the region that is now Syria and cultivated and stored grains used for brewing the beer served at humanity's first feasts. Feasts allowed the people of the settlements to bond, celebrate, and cultivate alliances with rival groups, and beer, as anyone who has ever gone to a party knows, lubricates conversation and lowers defenses.

This is the drink I still crave; the one placed in my hand within a few minutes of my arrival at a party that gave me something to focus on other than my own social anxiety and awkwardness, and closed the distance between strangers so I could ease my way into their conversation, the drink that helped me be wittier, more expansive, charismatic, and brave. People seemed to really like one-drink me. In truth, I miss one-drink me quite a lot.

One-drink humanity worked a little better than sober humanity during our transition from small bands of genetically interdependent people to larger, more complex civilizations. Picture the difference between those two phases of our development as a people as the difference between sitting down to a weeknight family dinner and attending a grand state dinner of one hundred guests and you are seated with North Korean supreme leader Kim Jong-un, your mother-in-law, George Clooney, Albert Einstein, your dentist, the president of the United States, and the girl who bullied you mercilessly in middle school. The former I could easily handle as now-sober Jess, but I get why the latter (in another life) would require at least a few preshow shots to calm my nerves.

As human settlements became increasingly complex and more socially and politically rigid, people began to feel a lot more angst, argues Jeffrey P. Kahn, psychologist and author of the book *Angst: Origins of Anxiety and Depression*. The rigid social structures of complex society dampen our human urge to explore, experiment, and express ourselves, so while those early dinner parties may have nourished the rise of civilization, they also triggered humanity's first cases of social anxiety. Kahn believes that alcohol provided the liquid courage needed to free us from our inhibitions and fuel our Epipaleolithic social jockeying, political debates, and power plays.

Alcohol allowed early humans to "break free from our biological herd imperative—or at least suppress our angst when we did," Kahn writes. It helped us violate social norms so we could rebel, explore, create, and collaborate in ways the pecking order of early societies may not have normally allowed. "Conversations around the campfire, no doubt, took on a new dimension: the painfully shy, their angst suddenly quelled, could now speak their minds."

That last line, from Kahn's *New York Times* op-ed, "How Beer Gave Us Civilization," describes the genesis of humanity's love affair with booze, down in the deepest roots of our collective family tree, but it's eerily reminiscent of the stories I hear in twelve-step meetings and read in my students' essays. "I felt different, like I didn't belong, as if everyone had been given the instructions for how to live but me. Then I had my first drink, and I found the solution, the thing that allowed me to be me, but better."

That's all humans have ever wanted to be: us, but better.

Alcohol has alleviated our collective angst, incited revolutionary zeal, and sustained cultural change for millennia. According to historian Gregg Smith, the birth and formative growth of our nation developed with beer. "Beer and society have been inseparable companions for thousands of years. Literally, the two have gone hand in hand," he writes.

The United States, after all, was born in taverns.

When the Puritans set sail for the New World, they carried three times as much beer as water, but by the time they arrived, they were nearly tapped out of both. The Pilgrims were used to drinking around a gallon of beer per person, per day, and as a result, everyone on the *Mayflower*, including the children, hovered around the modern legal limit for driving (rather, sailing) while intoxicated.

The colonists' first order of business upon landing in New England was to find a source of beer. They planted grains, built a tavern, and when the brewing wasn't going well, placed an ad in newspapers back home looking for a brewer willing to sail across the ocean to provide the colonists with their beloved English draughts. The preferred beer in England at the time was a hearty, dark ale that hovered around 6 percent alcohol and served as food, medicine, and comfort for the people who left their homes for the uncertain promise of a new land. Barley was in short supply on their side of the pond, however, so in the New World they fermented and distilled whatever they could get their hands on locally, including "pumpkins, and parsnips, and walnut-tree chips."

Brewing up local hooch wasn't so much about maintaining a buzz, though. There wasn't much else to drink and the colonists believed the cool, clear water of the Americas would kill them. This wasn't just a paranoid delusion: back home, Europeans used rivers, streams, and ponds as public septic systems, so their waterways were unpotable cesspools. The colonists could have boiled their drinking water, of course, but because they were living pre-Pasteur and his germ theory, the colonists had no way of knowing that beer was primarily safe to drink not because the malted barley, hops, or the bubbling fermentation imparted some sort of magic spell, but simply because it was boiled.

Aside from its health and safety benefits, beer provided emotional balm for the colonists' psychic wounds. Homesickness, mourning, struggle, and strife defined life in the New World. Fully half of the Plymouth colony was dead by the end of the winter of 1621, and the survivors' grief, guilt, and loneliness must have been intense. Once the colonists survived their challenging early years with the help of whatever ale-like drink they could brew, beer remained a source of daily hydration, sustenance, and the greater emotional health of the community.

Beer and distilled spirits such as rum and brandy lubricated the first wary meetings between the colonists and the indigenous tribes they encountered, although the historical record gets pretty hazy and biased around the topic of Native drinking. The historical record tends to be written by the victorious, educated White men, of course, and their accounts are heavy on stereotype, fearmongering, and myth. What we know for sure is that the introduction of alcohol was, and continues to be, devastating to the Native people of this country.

Here's just one example. *The Encyclopedia of New York City* posits the theory that the name "Manhattan" comes from the term *manachactanieck*, the "place of general inebriation," because in 1609 Henry Hudson offered the Native Munsee people alcohol to test their character and trustworthiness. Tolerance was low, initial enthusiasm for spirits was high, and addiction took over quickly. Even filtered through the European-centric, harshly anti-Native sentiments of the time, the destructive power of alcohol was clear: "Give two Savages two or three bottles of brandy. They will sit down and, without eating, will drink one after another until they have emptied them," wrote one missionary. The seeds of those early trades continue to bear poisonous fruit. Today, the rate of substance abuse among Native American adolescents is nearly three times higher than that of non-Native teens, the highest of any population group in the United States. Hudson's gift of booze on that island of general inebriation has led to the enslavement, subordination, and premature death of twenty generations (and counting) of Native people.

Tea tends to get credit for inciting revolutionary zeal and forcing the issue of war, but alcohol fueled the revolution. When the British sought to raise money for war by controlling and taxing all forms of alcohol, the colonists grew angry and flocked to their local taverns to blow off steam. Anger fueled more drinking, and drinking fueled early plans for revolt. The Boston Massacre began with an argument in a tavern. The promise of free beer lured men into consignment and hydrated their training days. The headquarters of the American troops at the Battle of Lexington were located in a tavern adjacent to the Lexington Green, and when the war was finally over and celebrations began, General Washington chose the Fraunces Tavern in New York for his final toast and farewell to arms.

Respectable Intoxication

Where drinking and drugging is concerned, adults lead and children follow. Adults set cultural standards for drugs and alcohol and for much of early American history, children drank alongside their parents. They were introduced to alcohol early, first as drops on the tongue, later in the form of dregs from an adult's cup or sitting next to a parent at the tavern and observing their behavior. One early American traveler wrote, "I have frequently seen Fathers wake their Child of a year old from a sound sleap [sic] to make it drink Rum, or Brandy." Another noted, "It is no uncommon thing to see a boy of twelve or fourteen years old . . . walk into a tavern in the forenoon to take a glass of brandy and bitters." According to Susan Cheever, author of Drinking in America, "Everyone drank, beginning at birth—infants were plied with rum to help with sleep—and ending at death." Kids drank beer or "flip," ale mixed with rum, raw egg, molasses, and sometimes chicken stock for breakfast. As a consequence, many kids spent their days nearly as tipsy as their parents.

Alcohol wasn't just for drinking; it was also good medicine. Colonial children drank variations on the ubiquitous ale and cider theme as supplemental nourishment and home remedies. "Cock Ale," a mixture of ale, raisins, spices, and chicken broth, was touted as a cure-all for everything from colds to the flu (and erectile dysfunction in dad), but for more severe injuries, they pulled out the hard stuff. Rum, gin, and brandy were thought to be capable of healing broken limbs, frostbite, shock, and snakebite. Parents dabbed alcohol on the gums of teething children and gave them hot toddies for colds ("toddies for toddlers," British philosopher John Locke called them), a remedy that lasted into the twentieth century, and into my own family. My mother remembers drinking them for her sore throats when she was a little girl. As time went on, and shipping routes increased, the availability of narcotic remedies did, too.

Drugs such as opiates have long been a part of medical remedies for both adults and children, easing the ache of common injuries as well as the pain of being human. Galen, the most famous physician in Western antiquity, recommended opium as a sleep aid, and the practice of using opium as a sedative persisted into the twentieth century. Women were instructed to paint their nipples or pacifiers with opiates to ease infants' teething pain and opiates were added to children's food to cure diarrhea. It was marketed as a "soother," in the form of sweet pastes, cordials, and nostrums in spite of the danger these "lethal lullabies" posed to infants. One of these remedies, the popular Mrs. Winslow's Soothing Syrup, claimed to "soothe any human or animal" and despite the American Medical Association's condemnation of the product as a "baby killer" in 1911, it remained on the shelves of corner drugstores until 1930. Physicians and pharmacists depended on the sale of narcotics and opposed legislation meant to curb its distribution. One pharmacist quipped, "If it were not for [opium] and my soda-water I might as well shut up shop."

However, once people realized the threat narcotics posed for children, legislation quickly followed. That's how drug laws happen, writes public health historian Alex Mold. "Indeed, fears about young people and drug use, both real and imagined, have helped to drive forward regulation of illicit drugs." Opiates passed out of cultural and legal favor with the passage of the Harrison Narcotics Tax Act of 1914, but another class of drugs was poised to fill the empty space on drugstore shelves.

Stimulants emerged as America's favorite little helper, first for adults, and later for children. And why not? Stimulants were sanctioned and distributed by the U.S. government, who used them to keep our military forces alert and awake. Even better, they were doctor-approved. "Many doctors believed that amphetamine and its derivatives had nearly unlimited potential. Meth[amphetamine] was considered a possible treatment for all manner of disorders, including epilepsy, Parkinson's, schizophrenia, and alcoholism—in addition to, of course, depression, obesity, and fatigue," writes Ryan Grim in *This Is Your Country on Drugs: The Secret History of Getting High in America.*

Beginning in the 1940s, amphetamines emerged as the solution to fatigue and boredom, an easy cure for weight loss, and an antidote for housewives' ennui. An advertisement in a 1950 physician's trade publication spelled out the drug's appeal: "Many of your patients—particularly housewives—are crushed under a load of dull, routine duties that leave them in a state of mental and emotional fatigue. Dexedrine will give them a feeling of energy and well-being, renewing their interest in life and living." Ads like this boosted cultural acceptance of the drug, and sales boomed. Amphetamine production quadrupled between 1949 and 1952, and by 1958, 3.5 billion amphetamine pills flooded the market. In 1967, less than a decade later, 8 billion amphetamine doses were prescribed, 80 percent of them to women. Physicians surveyed in 1962 revealed that half of their amphetamine prescriptions were written "for mental distress" and depending on who you asked, somewhere between one-half and two-thirds were prescribed for weight loss.

And then, in the fifties, a new potential drug market emerged for amphetamines: children. Pharmaceutical manufacturers began to advertise stimulants as a cure for "hyperkinetic" misbehavior in kids. Boys were more often identified as behaviorally problematic, and doctors pointed the finger at mothers. "Psychoanalytically-inclined child psychiatrists attributed [hyperkinetic misbehavior] to toxic motherhood, whereas the biologically inclined attributed it to 'minimal brain damage,'" writes Nicolas Rasmussen, historian and author of *On Speed*. The condition was considered so rare that an expert testifying before Congress in the late 1960s predicted an annual supply of a few thousand doses would be more than sufficient. That expert's prediction fell just a wee bit short. In 2016, 48.18 *metric tons* of the four major stimulants used to treat attention deficit hyperactivity disorder (ADHD) were prescribed in the United States.

After a decade of taking drugs in order to conform to society's gender and behavioral expectations, it's no wonder that adolescents of the 1960s and 1970s favored substances like marijuana and LSD, substances that allowed for rebellion, escape, and freedom. Like early humans who drank beer to ease their angst, many adolescents drink and take drugs to exert their newfound independence, separate from their family of origin and its authority figures, and to find their place in society.

In the 1960s and 1970s, the drug culture offered both: separation from the relatively conservative politics of generations past and an invitation to take part in a new order of rebellion and change. Kids could figure out who they were free from rules and conventions, and as drugs and alcohol shrugged off their mantle of criminal deviance, they took on a new allure among young rebels determined to push back against their parents, government, wartime brutality, and the strict morality of generations past. They wanted to change the world and drugs helped them envision that gentler, softer, more spiritually expansive world. Writers like Jack Kerouac and Allen Ginsberg romanticized drugs as a way to that place, as a portal into the subconscious and a way to express one's deepest creative ideas and impulses (although even inveterate drug enthusiast Ginsberg drew the line at amphetamine use, calling it a "plague in the whole dope industry"). LSD and marijuana moved into the upper class and gained popularity for their ability to expand consciousness and sensory perception. Acceptance and promotion of drug culture in the media spread the perception that "everyone" did drugs, which normalized drug use and encouraged new users to join in the rebellious fun. This normalization, however, is also what prompted the federal government to declare war.

In 1971, President Nixon announced drugs were "public enemy number one" and launched an "all-out offensive" against both the supply and demand sides of the drug trade. Nixon requested $115 million for the effort, and allocated a significant portion of that money to drug abuse treatment and prevention. This funding, the first of its kind, was well timed, as thousands of soldiers were about to return home from Vietnam addicted to heroin. Unfortunately,

Nixon's federal drug treatment programs were short-lived, as Ronald Reagan cut them out of the federal budget to make room for his wife's highly publicized yet ineffective "Just Say No" policy of the 1980s. Reagan's timing was disastrous, as crack was about to take hold in urban America and Nixon's prevention funds might have slowed that epidemic.

Nancy Reagan wasn't the first to impose her ill-conceived notions of moral temperance on the country, and given that another first lady, Betty Ford, had recently publicized her own inability to say no to drugs and alcohol, Nancy's message was particularly tone- (and science) deaf. But Betty Ford's honesty about her addiction to drugs and alcohol, paired with Nancy's misguided yet well-intended slogan, birthed a new recovery movement in this country.

The temperance movement, and by extension the recovery movement, had been around for 170 years by the time former first lady Betty Ford went public with her alcoholism. When the first temperance society was born in 1808 in Moreau, New York, the word "temperance" meant different things to different people. Some opted for a so-called long pledge to abstain from all alcohol, including beer, wine, and cider. Others, however, went for a "short pledge," which forbade only hard liquor, while keeping beer, wine, and cider on the table, so to speak. By 1835, two million people had signed temperance pledges, but disagreements over the meaning of sobriety began to fracture groups within the movement.

Despite these rifts, consumption of distilled spirits fell precipitously between 1830 and 1840, and states began passing temperance legislation, starting with Maine in 1851. Unfortunately, bans on the sale of alcohol were toothless, and the illegal sale of alcoholic beverages carried on publicly and openly. Hannah Jumper became so fed up with the unbridled, raucous drinking among the men of Rockport, Massachusetts, that she led her hatchet gang of two hundred pissed-off women in a fearsome and thorough "likker raid" in July 1856. Over the course of five hours, they smashed every barrel, jug, cask, and bottle of booze in town. Her sister in sobriety, Carry Nation, known as the "Kansas Cyclone," led the charge for sobriety out west, where her fans would come out to cheer her along, and purchase the Carry Nation swag—pewter replicas of her hatchet and the bashed-up remnants of whiskey kegs—that financed her crusade. By 1878, the largely female temperance movement drove per capita consumption of alcohol down to eight gallons a year, and Nation died in 1911 believing that a national prohibition was within reach.

In name at least, she was right.

Unfortunately for Hannah and her sisters, Prohibition did not reduce America's alcohol intake. Almost as soon as it began, brewing supply stores proliferated, and Americans returned to their colonial do-it-yourself roots and cooked up whatever produce or grain they had on hand to make beer, moonshine, bathtub gin, and whiskey in their home stills. Taverns and bars closed, but illegal speakeasies opened in their place, and by the end of Prohibition, the number of speakeasies in New York City was double that of the pre-Prohibition legal establishments.

Legal or no, culturally accepted or no, Americans continued to drink.

For all its failures, Prohibition did affect one shift in America's views on substance abuse. Before Prohibition, the problem of substance abuse was attributed to saloons, bars, and taverns (cf. Hannah and Carry) and so, the thinking went, eliminate the supplier and you eliminate the problem. When that approach failed, blame shifted to the alcoholic.

The Invention of the Addict

For much of American history, drunkenness was assumed to be voluntary, that people drank because they wanted to, and all drunkards needed to do was just say no. Blaming the drunk for their overindulgence is a long-standing tradition in America. Drink too much in colonial New England and you were likely to end up in the village stocks for a few hours and fined five shillings to boot. The crime of overindulgence was not only a rebuke to the God who provided alcohol as a gift to mankind, but it exposed a dangerous failure of man's morals and will.

The first physician to call attention to the emerging health dangers associated with excessive alcohol consumption and suggest that the alcoholic was not entirely to blame was Dr. Benjamin Rush. Rush was the first to use the word "disease" to describe alcohol and its effect in his 1791 article, "An Inquiry into the Effects of Spirituous Liquors on the Human Body and Their Influence upon the Happiness of Society," and debate over the appropriateness of the word "disease" in reference to substance abuse continues today.

The language we use to describe, define, and categorize substance use and abuse matters because it drives cultural attitudes around addiction, funding for future research and insurance company policies that determine coverage, and ultimately, whether a child in Alabama or Montana or Vermont can find and afford treatment.

Proper naming is especially important when talking to children. It matters what we call things, or whether we call them anything at all, for that matter. When I was young, and my relatives were drunk or zoned out on pills, I was told to be quiet, that he was "taking a nap" or she "has a headache." There's a place for linguistic hedges when talking to very young children, but when we tell kids they are not seeing or hearing what they *know* they are seeing and hearing, that's called gaslighting, an emotional manipulation tactic used to gain power over people and ultimately make them question their own sanity. Susan Cheever, daughter of writer and alcoholic John Cheever, writes of her father's drinking, "The recovery people say that alcoholism is like an elephant in the living room, and that living with alcoholism induces insanity because everyone has to pretend that the elephant isn't there. Pretending that things are not as they seem—that you don't see what you do see, that you don't hear what you do hear—makes children crazy." The first step in not repeating the mistakes of generations past is to know and use honest language, whether it's an elephant or mommy's inability to stop drinking wine.

The American Psychiatric Association (APA) has spent a lot of time and effort on proper naming. Addiction is formally defined by the APA's *Diagnostic and Statistical Manual* (*DSM*), and with each of its new iterations, the debate over what is—or is not—a psychiatric disorder evolves. In the 1980s, when the *DSM-III* was revised, there was much discussion about whether to adopt the term "dependence" or "addiction." Those in favor of the term "addiction" sought to differentiate the clinical condition of compulsive drug taking from the state of physical dependence on a substance, which can happen to anyone and be completely normal. Dependence is too broad a term to characterize the condition of addiction, detractors argued. A vote was taken at one of the last meetings of the committee, and "the word 'dependence' won over 'addiction' by a single vote." The *DSM-IV* adopted the word "dependence," and the debate dragged on into meetings planning for the manual's fifth edition, which combined the terms "substance abuse" and "substance dependence" into one broad category, termed "substance use disorder."

Language meant to describe the substance user has evolved, too. Disorder-oriented labels such as "addict," "junkie," and "alcoholic" are passing out of favor, while person-oriented language such as "person with substance abuse disorder" has become the preferred (if unwieldy) term. The 2017 edition of *The Associated Press Stylebook* codified this linguistic evolution for journalists, declaring that the noun "addict," and by extension "alcoholic," should no longer be used to describe a person with a substance use disorder.

There's lots of support for this linguistic and philosophical approach. The authors of two recent books about women and drinking, Holly Whitaker (*Quit Like a Woman: The Radical Choice to Not Drink in a Culture Obsessed with Alcohol*) and Laura McKowen (*We Are the Luckiest: The Surprising Magic of a Sober Life*), argue for abandoning the word "alcoholic" altogether. Alcohol is a poison: a toxic, corrosive chemical that alters our mental state and damages our bodies on the cellular level from the moment it passes our lips, Whitaker and McKowen argue. Alcohol, after all, is ethanol, "the same thing we use to make rocket fuel, house paint, antiseptics, solvents, perfumes, and deodorants, and to denature (i.e., take away the natural properties of, or *kill*) living organisms," Whitaker writes. When we normalize drinking for the masses while marginalizing the unfortunate few who can't handle their drink, we shift the stigma away from the toxic chemical and onto the people who get sick when they drink it.

In my talks or written work, I use descriptors that focus on the substance, not the person, yet I still call myself an alcoholic. The word is clear, unequivocal, honest, and blunt, and defines one of the most important chapters of my history.

My students always ask if I have a favorite page or image in Jarrett Krosoczka's *Hey, Kiddo*, and of course I do. It is in the chapter called "Disclosure." Jarrett's grandparents sit him down on the couch so they can tell him the truth: his mother has not been around for a while because she is in jail as a result of her heroin addiction. Jarrett cries and clings to his grandfather for comfort while his grandmother pats his back. Just under the image, Krosoczka writes, "I knew in that moment, when my grandfather told me the plain truth, that life wouldn't be the same for me. It didn't change the circumstances, but it shifted my perspective."

That's all I can hope to do for my own children and my students, to shift their perspective. They will feel angst as they grow up and their lives become more complicated. They may even yearn for escape—from their stress, from their sadness, from themselves. But if we can help our children understand the reasons they may be tempted to drink or do drugs on their first day, they may never end up having to recount the story of their worst day.

CHAPTER 3

WIRED FOR RISK

A Primer on the Adolescent Brain

A lot has changed in the science of adolescence since the late 1990s, much of it thanks to the invention of functional magnetic resonance imaging (fMRI). Until the fMRI came along, and scientists were able to look inside the brain and witness its functioning in real time, science operated on the false premise that because the human brain reaches its adult size by age ten, it is fully developed by then as well.

Now, I've taught ten-year-old children, and even without an fMRI, I can attest that their adult-sized brains are years away from being fully cooked. The lower, more primitive regions of the brain that allow them to perceive their environment and react to it work brilliantly, and often in overdrive. Ten-year-olds are excellent perceivers and reactors. The higher, more evolved parts of their brains that temper all this perceiving and reacting are not up and running yet, which explains why the perception "she took my toy" results in an unfiltered, unrestrained reaction: a smack to the arm or pull of the hair.

Once kids hit adolescence, however, the other, more rational and reasonable areas of the brain begin to catch up to the perceiving, reacting parts. Around the time a kid enters puberty, the adolescent brain is in the second of two intensive phases of growth and development. The first, from birth to about age three, is the one that gets the most positive press, what with the easily charted, adorable milestones. Baby's first smile, first word, and first steps are tangible, social media–ready landmarks—proof that serious neurological growth is happening in her milky-soft, sweet-smelling head. Adolescents are different, and not just from an olfactory perspective. Their milestones are less obvious to the outside observer. Have faith. Just as surely as that infant pulled herself up to her feet and started toddling about, teenagers, too, will find their footing.

Between ten and twenty-five, her brain will mature rapidly and will be more sensitive to the great potential and perilous dangers of her environment. Her propensity and susceptibility to substance use depends on the course of her neurological development, and in turn, the course of her neurological development is influenced by her substance use.

In order to prevent adolescent substance abuse and the harm it can do to her development, we have to understand her brain, so that's where we will start. Once we understand what's going on within her adolescent brain, we can begin to understand why she is at greater risk of substance use during this period and what these substances do to her neurological and cognitive growth.

A Primer on the Adolescent Brain

Some animals are born with the ability to get to their feet and keep up with the herd just moments after birth, but not humans. As compared to other animal species, humans are born underdeveloped and helpless in an evolutionary trade-off. In exchange for our big brains (read: big heads) and upright stance (read: narrow pelvis), human babies must be delivered when their brains are less than 30 percent of their adult size or they'd never fit through the birth canal. For a human baby to have the same neurological and cognitive development as a newborn chimpanzee, human mothers would have to gestate babies for 18 to 21 months. As the mother of two large-headed boys, I am happy to play along with this evolutionary bargain.

A newborn baby arrives in the world with little more brain function than our earliest ancestors had when they hauled their fishy bodies out of the ocean. Infants are born with the capacity to maintain basic life functions (breathe, eat, pee, poop, sleep, and keep the heart thumpa-thumpa-thumping) and the instinct to interact with the people she depends upon for survival, people who will jump-start her cognitive development with their eye contact, talking, and cuddling.

She's already got most of the one hundred billion brain cells (**neurons**) she'll have in her lifetime, but they can't communicate with each other very well yet. Those connections will develop over the first twenty-five years of her life, beginning with the most primitive structures in the lower brain and finishing in the most highly evolved, complex structures in the top and front that will grant her the higher-order organizing, planning, reasoning, and impulse-control abilities she'll need in adulthood.

Before her first birthday, an infant's brain doubles in size and

creates 100,000 new **synapses** (connection points between neurons) per second. Eventually, she'll have a quadrillion synapses, but
for now, she has precious few, and the ones she does have are located in the base of her brain. For the time being, her lower brain
runs the show.

Her lower brain is made up of a **brainstem** that sits at the
intersection of her brain and spinal cord and manages her basic
life functions. Her **basal ganglia** control voluntary movement and
will, over the course of her first year, grant her the ability to control her hands and arms, move food to her mouth, grab her toes,
roll over, crawl, and eventually walk. She will learn procedures,
habits, and associations between her behavior and rewards in her
nucleus accumbens. The support system for her emotions, long-
term memory, motivation, drives, and instincts such as hunger, sex,
and parenting lies in the **limbic system**, which is made up of the
hippocampus, where the bulk of memory processing happens, and
the **amygdala**, a structure vital for interpreting social cues, such as
facial expressions, eye gaze, social hierarchies, peer pressure, reputation, and physical appearance.

It is here, in the limbic system, where she launches reactions to
her environment and creates her most vivid, emotionally charged
memories. If she needs to freeze, fight, or flee, this system works
great. It's not much help, though, when it comes to the business
of adulting.

For the higher-order thinking, the planning, prioritizing, goal-
setting, strategizing, or weighing a complex or delayed risk-reward
ratio, she will need a **frontal** and **prefrontal cortex**, structures that
won't be fully wired up, internally or to the rest of the brain, until
her mid-twenties. The process of connecting the frontal lobe and
becoming cognitively dependent on the more adult, evolved parts
of her brain is called **frontalization**, and it takes about fifteen years,

an interval that, for a parent patiently waiting for her teen to grow some reason and common sense, can feel like forever.

Gap Years

When a teen does something impulsive or foolish, with no fore-thought or regard for consequences, and we ask, "Why didn't you *think* before you acted?" we are tilting at the windmills of adoles-cence, weeping and wailing over behaviors they are not yet able to fully control. Teens are more focused on rewards than adults, and less able to weigh the long-term consequences of their actions. Neuroscientist Aaron White of the National Institute on Alcohol Abuse and Alcoholism puts it this way: "Adolescence is a develop-mental period characterized by suboptimal decisions and actions that are associated with an increased incidence of unintentional injuries, violence, substance abuse, unintended pregnancy, and sex-ually transmitted diseases." Scientifically speaking, teens are capa-ble of some truly stupid shit.

However, even in the midst of our most frightening parenting moments, when they make us cry with frustration, when they do foolish, dangerous things and show little concern for their safety, remember that this is all developmentally appropriate. They are in their gap years, a trying period in which the system that drives their need for independence, risk, new experiences, and sensations is in high gear, but the system that drives higher-order planning, caution, and rational thought is still in park. It would be lovely if these two systems evolved on the same timeline, but they do not, and the gap between the mature, highly active limbic system and the immature, less than fully functional prefrontal cortex is what gets kids into so much trouble and drives teens' vulnerability for substance abuse.

There's a Great Future in Plastics

The adolescent brain may be frustrating, and slower to mature than we may like, but when I'm at my wits' end with my kids or my students, I focus on the spot right between their eyes, where the prefrontal cortex resides, and remind myself that a small miracle is taking place in there.

Need tangible proof? Try taking some kind of class with your kid, something that requires cognitive dexterity and mental flexibility. I took acoustic guitar lessons with my son Ben when he was fourteen and I was forty-five, and things got humbling fast. He'd acquired an entire classic rock repertoire by the time I could reliably stumble through a simplified, three-chord rendition of Fleetwood Mac's "Second Hand News."

Ben learned guitar faster than I did, not (necessarily) because he's smarter than I am (he is), but because infant and adolescent brains are extremely **plastic**. Plastic brains learn quickly, but they are also highly sensitive to the good and bad in their environment. This is a boon for learning, but it is also a dangerous, precarious time. During this transition from child to adult, adolescents are much more sensitive to negative environmental influences such as trauma, stress, social rejection, and sleep deprivation. "Plasticity is the process through which the outside world gets inside us and changes us," writes adolescent psychologist Laurence Steinberg in his book *Age of Opportunity*. In other words, before your teen can strike out on her own to change the world, the world will change her.

Plasticity happens through two processes: **synaptogenesis** and **myelination**. Infants and young children may be bursting with neurons, but communication between those cells is weak and inefficient if it exists at all. It's not as if the space behind their foreheads is empty; the frontal lobe and its neurons have been there since birth. They're just not wired up yet to the rest of the brain. Communication between brain cells happens via synapses, and not all of her neurons are equipped with them yet. In her first couple of years of life, she created up to two million synapses a second, mainly in the lower, primitive brain structures she needed for all the pooping, peeing, breathing, and sleeping. Now that she's in adolescence, she can start building them in the more-evolved areas of the brain, concluding with the bit just behind the forehead, the prefrontal cortex. In order to stay alive, brain cells must stay active, make new connections, and maintain communications with other cells. When communication shuts down, as it often does in the presence of addictive substances, entire pathways can die off in a loss of potential that can be catastrophic for the adolescent brain.

When allowed to progress unimpeded, and with plenty of positive environmental stimulation to promote growth, synapses proliferate at an incredible rate in a gorgeous pattern that looks a lot like the spreading of a rain forest canopy, trees branching out into two, then four, then eight in a process called, aptly enough, **arborization**. While arborization is going on, and she's making connections here, there, and everywhere, her neurons are also becoming more efficient through a separate process called myelination, in which **oligodendrocytes** insulate her neurons with a fatty sheath called **myelin**. Myelin is white, so it turns her gray matter (uncoated neurons) into white matter (coated neurons). Gray matter is a poor conductor, much like bare, copper electrical wires without their coating. If we were to plug those in, they would short out and misfire in all directions. Once coated with myelin, however, nerve impulses travel faster, more of the impulse arrives at its destination intact, and a higher volume of signals can travel over brain cells in more rapid sequence with decreased recovery time between each impulse. Better conductivity allows for more efficient thinking, creation of memories, and dependable retrieval of those memories.

Synaptogenesis and myelination take place over years. A kid may be capable and competent one day, and a total, catastrophic mess the next. She will be perfectly able to reason in a rational and mature fashion in first-period science class, but by the end of the school day, she may devolve into a weepy, frustrated mess. Adolescent brain development is messy and imperfect when viewed day to day, but in the bigger picture, progress is being made. Just step back a little. Be patient with the short-term outages and be grateful for what's functional on any given day.

Meanwhile, back in our metaphorical rain forest canopy, the final phase of development is taking place, which is good, because things are getting loud and chaotic. She's been synaptogenesis-ing and myelinating for so long, her brain has become dense and thick with redundant neural connections. Each neuron can send up to a thousand impulses every second over ten thousand synapses, but with that many connections, unnecessary replication of effort begins to mess with her cognitive efficiency.

Beginning in about mid-adolescence, the brain begins to cut back the dense canopy of connections through a process called **pruning**. Superfluous and redundant neural pathways are removed in favor of fewer, more efficient connections. Sure, some gray matter is lost along the way, and if we were talking about an elderly person it would be cause for concern, but because loss of gray matter in adolescence is accompanied by dramatic increases in white matter, it's all good. Teens end up with a net gain in cognitive speed and flexibility.

By her mid-twenties, she will be at peak efficiency. She will have synapses galore. Her gray matter will have been converted to white where necessary, and she will be fully equipped for the tasks of adulting. Forget *quinceañera*, forget sweet sixteen, casting a first vote, or toasting a first legal beer. The milestone of cognitive maturation is the one worth celebrating. If she begins drinking or taking drugs in her mid-twenties, once her brain development is complete, bad things can still happen but she is at *much* lower risk of brain damage and poor mental health outcomes, including substance use disorder.

The simple solution, of course, is to keep kids off drugs and alcohol until they are twenty-five, right? I mean, how hard can that be? Harder than we adults can imagine, unfortunately. The adolescent brain is uniquely wired to crave the very experiences drugs offer, including risk, escape, novel sensations, reduced anxiety, and peer acceptance. The whole point of adolescence is to strike out on your own, prove yourself in a competitive world, establish your own unique identity, and forge a path into adulthood, and the scary truth is that kids who actively pursue new, daring, slightly scary experiences are more likely to accomplish these goals. The trick is to encourage and channel the risk-taking and sensation-seeking into healthy directions while helping them manage their impulsivity and appetite for dangerous risk. But first, we have to understand dopamine.

My Chemical Romance

Until the frontal lobe comes online in the late teens or early twenties, the limbic system is large and in charge, and it runs on chemicals called neurotransmitters. There are many, but we will focus on dopamine. Dopamine is our chemical reward, our reason for getting out of bed in the morning. It drives us to hunt down food, sex, exciting experiences, and, yes, drugs and alcohol. Dopamine, at its essence, *is* drive.

Adolescents have lower baseline levels of dopamine than young children and adults. It may seem counterintuitive but it makes sense: low baseline dopamine levels make life seem pretty boring. Weekends with nothing to do? Boring. Activities they used to love when they were little? Yawn. That summer vacation they looked forward to all year? OMG, so boring.

While baseline levels are low, teens' response to dopamine release is much more acute than in adult brains, and nothing releases dopamine like risky exploits that provide new sensations and experiences. Well, except drugs, but we will get to that in a second. When the things they used to love like reading, playing in a tree house or on the swings, or simply hanging out with friends do not provide the dopamine hit they used to, it's natural that they would seek out thrilling, risky, dangerous, and scary experiences that will.

Heightened dopamine response is what makes adolescence so exciting and visceral and memorable. The teen hippocampus, where short-term memories are archived for long-term retrieval, is working in overdrive. The limbic system is running the show, remember? The hippocampus is especially good at translating emotional, exciting experiences into memory; hence, teen love is more thrilling, heartbreak more devastating, anxiety more overwhelming, pleasure more all-encompassing.

Take a second to close your eyes and think back on your first kiss, first love, greatest teen accomplishment, scariest adolescent memory. See what I mean? I can hardly remember what I talked about last night over dinner, but I remember every detail of my first kiss. I remember where I was (the basement of his house), what I wore (a matching skirt and top from The Limited), what song was playing ("Open Arms" by Journey, of course), and my inability to breathe for a few moments after.

I remember the moments leading up to the kiss because dopamine had started to flow well before it happened (I knew it was coming as kids were pairing off, and did I mention "Open Arms" was playing?) but at the moment of contact, *whoosh*. Huge dopamine rush. The moment the kiss was over, however, my dopamine levels began to plummet back to baseline or below, depending on the reason for the release, triggering a need to find something to stimulate my next hit. You can see how this cycle can get kids in trouble. The more powerful and robust our dopamine cycle of low to high to low, the more likely we are to become dependent on whatever caused the rush, to pursue the next high and eradicate the lows.

Kissing is great and all, but even adolescent kisses can't match the dopamine rush of drugs and alcohol. They hijack the brain's reward system, flooding dopamine receptors and causing levels to rise higher than they could ever go on kisses or roller coasters. The dopamine rush of drugs is followed by a precipitous fall to levels below baseline, because as the body struggles to keep some kind of balance in the face of this dopamine flood, it releases less dopamine, lowering a teen's baseline levels even further. Life between hits becomes even more boring, and eventually, as Dr. Corey Waller of the American Society of Addiction Medicine explains, "You can't produce enough dopamine to get out of bed." In other words, kids who use drugs and alcohol early in adolescence rewire their brains such that the substance is required in order to feel normal. Not high. Not buzzed. Normal.

The work of the adolescent brain is to develop and maintain a moving target of "normal" as it matures, and it simply can't do that when it's competing with addictive substances. Keeping kids off drugs and alcohol is difficult enough, but as those substances hijack, then completely take over the reward system in the brain, it only gets more challenging.

Our best and most powerful weapon, according to research and lessons learned from years of failed drug prevention strategies, is transparency, honesty, and evidence-based information. I discuss the folly of lies and scare tactics fully in chapter 9, but for now, the more kids (and the adults in their lives) understand about how drugs and alcohol impact the adolescent brain, the more opportunities we have to combat misunderstandings and opinions about the hypothetical "everyone" who is doing it, how much they are doing, and how it can permanently alter their lives.

What follows are the facts about what substances do (and don't do) to the adolescent brain.

Youth Is Wasted on the Young

The brain plasticity of adolescence is what allows for massive learning and growth, but when the chemical complication of drugs and alcohol enter the mix, they can interrupt and even permanently derail brain maturation. There are many different classes of drugs, and each has a slightly different effect on the brain (I will go through each at the end of this chapter), but taken as a whole, drug and alcohol use is a disaster for learning.

Learning, as any teacher knows, does not happen because of any one thing. It's a complex interplay of attention, cognitive processing, working memory, visuospatial processing, information retrieval, memory formation, verbal learning, goal-directed behavior, and maturation of executive function skills. Every one of these factors is vital, and addictive substances disrupt them all.

Research is emerging about how, exactly, substances mess with the teen brain, but evidence points to drugs' interference with the way neurotransmitters bond to brain cells, communicate their electrical signals, and determine whether the neuron fires (an **excitatory state**) or not (an **inhibitory state**). Some drugs increase the excitatory state, others the inhibitory state, but interfering with either one throws off the delicate, constantly changing balance between on and off, high and low, excited and inhibited. There are more than one hundred different kinds of neurotransmitters in the brain, and they can join forces to create an infinite combination of excitatory and inhibitory states in the brain. Clear, unimpeded communication between brain cells is critical because neurons must have other neurons to talk to in order to stay alive. In an adult brain, a communication pathway between neurons may be able to shut down for a while, then recover once the drug is out of the body, but in a developing adolescent brain, where neurons must connect in order to continue maturing, an entire pathway could die off in a massive, permanent loss of cognitive potential.

Some of the most dramatic damage to the brain is caused by drugs that cause sedation and reduce anxiety. These drugs include alcohol, benzodiazepines, and marijuana because their tendency in the brain is to inhibit memory formation and retention. Unfortunately, drugs that cause sedation and reduce anxiety are also the drugs kids seek out first and use most often, and once they begin using these drugs, their lives can become a self-perpetuating cycle of despair. A kid who has academic problems is at higher risk for substance abuse, and kids who use addictive substances are at higher risk for developing academic problems. And so it goes, and so it goes.

Finally, substance use increases already elevated stress levels in adolescence. The stress response in an adolescent brain is more easily triggered and remains activated longer. As a consequence, stress feels more stressful (remember, emotional centers in the lower brain are working in overdrive, and the regulatory system of the frontal lobe can't manage impulsivity and emotion yet), so memories of stressful events are more vivid and life feels more overwhelming and unpredictable to an adolescent than it does to an adult. While some stress is good for brain development, too much changes the amygdala on a cellular level and causes it to become hypersensitized to perceived threats. Kids often drink (and use drugs like marijuana and opiates) to silence their anxiety and stress, but these substances can increase stress levels in the long term due to these changes in the amygdala and because of the increased complexity of juggling substances with the business of everyday life.

Increased stress levels are exhausting, but unfortunately, drugs and alcohol interrupt sleep cycles and duration even when they facilitate falling asleep. Stress and anxiety cause teens to use more substances, and using more substances further increases stress levels, which further undermines sleep quality and duration. The quickest way to interrupt learning in the brain is to introduce high levels of stress, so you can see where this is going. Exhausted, stressed-out teens can't learn as well as their sober peers, which can lead to academic failure. Academic failure feeds substance use . . . and here we are again in that maddening, self-perpetuating loop of substance abuse.

Now let's look at what different addictive substances do to the adolescent brain. Some of the studies I'm going to mention use animal models (rats!) because rat brains function a lot like ours when exposed to addictive substances, and because it would be a sticky ethical and legal wicket to expose adolescents to drugs in the name of scientific discovery. I will be sure to point out the instances of animal, rather than human, subjects as they come up.

It's also worth noting that when studying substance use in adolescents, researchers tend to rely on self-reporting, which raises its own statistical issues and serves as a great reminder to anyone who works with young people. Just as it can be difficult to elicit details about what happened at school today ("Nothing," being a popular teenage response), getting accurate information about substance use is a real statistical challenge. What's worse, when kids don't trust the adults in their lives, they are likely to hold back and underreport involvement in risky or illegal activity. This is also why long-term, trusting relationships with parents, health care professionals, teachers, and mental health counselors are vital not only for research purposes, but to ensure that your child receives continued, developmentally appropriate screening for drug and alcohol use, abuse, and dependency.

Alcohol: I'm Wasted and I Can't Find My Way Home

There's a lot of scary information in the media about kids and substance abuse, so let's start with some good news: rates of alcohol use during adolescence have been falling or staying the same for the last generation or so. Currently, 59 percent of high school students report having had at least one drink ("more than a few sips") by the time they graduate, and 24 percent have done so by eighth grade. However, adolescents consume alcohol differently than adults. When kids drink, they tend to binge drink, and binge drinking is associated with increased rates of smoking, illicit drug use, drunk driving, risky sexual behaviors, auto accidents, physical fighting, fewer hours of sleep, and lower grades. Currently, binge drinking is defined as four or more drinks in succession for women and five or more drinks in succession for men. I say "currently," because these will change if the American Academy of Pediatrics (AAP) has anything to say about it. The AAP recommends the definition be reduced to three or more drinks for children ages nine to thirteen and girls under seventeen, four or more drinks for boys fourteen to fifteen, and five or more drinks for boys sixteen and over to compensate for adolescents' smaller and less developed body size. By the AAP standard, rates of binge drinking among adolescents would rise, but currently, "Among youth who drink, the proportion who drink heavily is higher than among adult drinkers, increasing from approximately 50% in those 12 to 14 years of age to 72% among those 18 to 20 years of age."

Adolescents don't just drink differently than adults; their bodies process it differently, too. Teens are more likely to feel the positive sensations associated with drinking, like loosened inhibitions and decreased social anxiety, but less apt to experience the negative ones like sedation (rats!), loss of motor control (rats!), and hangovers (also rats, which raises the question: How do you know if a rat has a hangover?). They feel brave, loose, social, and expansive, but the usual checks on drinking to dangerous excess, such as a subjective feeling of being "too drunk" or passing out, are less likely to kick in. Because teens don't feel as drunk or as tired as an adult might given the same blood alcohol content (BAC), they may believe they are okay to drive, expose themselves to social or sexual risk, or drink to the point of alcohol overdose.

Girls are at higher risk when it comes to many of these dangerous outcomes because their bodies tend to be smaller, with a higher percentage of fat, which means they can't process the same amount of alcohol as boys can. They also produce less of an enzyme required to break alcohol down in the stomach, which means more alcohol makes it to their bloodstream. A girl's BAC may be up to 30 percent higher than a boy's, even if they are the same age and drank the same amount.

The brains of kids who drink are different from the brains of sober kids in both form and function. Alcohol disrupts the formation of new memories by inhibiting brain activity generally (it's a central nervous system depressant) and by interfering with the hippocampus specifically. Unlike other areas of the brain, the hippocampus continues to create new neurons, cells that are absolutely critical for memory formation, and alcohol interrupts this process. Kids who drink tend to have smaller hippocampi, and this effect is age-dependent, meaning the younger she is when she starts drinking, the smaller her hippocampus will be. Memories are formed here, so a reduction in hippocampus size may be one reason adolescents are more prone to suffer from blackouts, and why this tendency persists into adulthood. Blackouts (or brownouts or grayouts, all variations on this same theme) are scary because entire chunks of time—experiences, conversations, thoughts—disappear. It's not that these memories can't be retrieved; it's that they were never created. We take injury-related memory loss seriously, so why not alcohol-related loss? No matter how it happens, memory impairment is a sign of brain damage.

As many as 40 percent of college students report having experienced a blackout, and those who black out more frequently do so at significantly lower blood alcohol concentrations than their peers. This is called the **tolerance effect** and it's something I talk about quite a bit with my own kids. If kids black out early on in their drinking, especially after drinking less than their peers, it can be a flashing red warning sign that they are at greater risk of alcohol dependence during their lifetime. I don't tell my kids this to scare them, but to arm them with information. There's a joke among people in twelve-step recovery that using is never as fun after you've been to a few meetings because you have too much information. All that pesky knowledge gets in your head and harshes your buzz.

Damage to the adolescent brain can extend beyond the hippocampus into the delicate, still-developing cortex. Alcohol can kill off gray matter in the frontal lobe, and as I've mentioned, teens need as much frontal lobe as they can get. In one large, long-term study, teens who admitted to drinking lost so much gray matter in their frontal lobe that the lead researcher, psychologist Edith Sullivan, described the loss as "striking." Frontal lobe gray matter loss leads to impairment in decision making, judgment, inhibition, and can lead to inappropriate social behavior and risk-taking. Damage to the frontal lobe isn't reserved for problem drinkers, either. Chronic drinkers whose consumption and habits don't rise to the level of dependence can have it, too, and it can persist even with long-term abstinence.

Alcohol damages developing brains, whether during gestation, infancy, or adolescence. According to the Centers for Disease Control and Prevention (CDC), there's no safe amount of alcohol during pregnancy. Once a baby is born, the CDC advises mothers to avoid breastfeeding within two hours of consuming a drink because alcohol can be detected in breast milk for two to three hours after consumption. If adolescence is second only to infancy in terms of brain development, it makes sense that even small amounts of alcohol can damage the adolescent brain. In fact, the authors of a massive, international study of global disease burden go even further and advocate for total abstinence from alcohol in the general population. "Our results show that the safest level of drinking is none." Given these warnings, it's time to come to terms with the reality that drinking during adolescence is not harmless and shift our parenting strategy away from blind eyes and harm reduction, and toward zero tolerance.

Nicotine

Nicotine use, unlike the use of other drugs, is increasing thanks to the popularity of e-cigarettes. In 2018, three million kids in the United States, or 21 percent of high school students, used e-cigarettes, and 28 percent of those kids do it at least twenty times a month. The number of kids using e-cigarettes was 38 percent higher than the year before, despite the fact that cigarette use continues to fall. In a statement released by the CDC about these figures, director Robert Redfield said, "The skyrocketing growth of young people's e-cigarette use over the past year threatens to erase progress made in reducing youth tobacco use. It's putting a new generation at risk for nicotine addiction."

Nicotine is highly addictive because it delivers immediate cognitive rewards, including dopamine release, increased focus and concentration, reduction in anxiety and hyperactivity, and appetite suppression, and the rituals around nicotine use can be as comforting and pleasurable to users as the nicotine itself. The physical damage tobacco wreaks on the body gets a lot of publicity, but we still don't fully understand what nicotine does to the adolescent brain. What we do know is that kids who use e-cigarettes are more likely to start smoking traditional cigarettes, that young smokers develop nicotine dependence and tolerance more rapidly, and that teens who smoke are twice as likely as their nonsmoking peers to suffer from depression. The kicker on the depression statistic is that depression during adolescence leads to higher rates of smoking, so once again, we find ourselves flailing about in that dizzying, self-perpetuating loop of cause and effect.

Despite news of several thousand cases of acute lung disease and at least sixty deaths due to the presence of vitamin E acetate in unregulated vaping products, kids are still using nicotine e-cigarettes. Many of the kids I talk to about their nicotine use point out that the chemical that was making people sick (vitamin E acetate) isn't allowed in vaping products anymore, so there's nothing to worry about. That's not quite true. While e-cigarettes are relatively less toxic than traditional cigarettes in that they don't contain tar and other carcinogens, they significantly increase the risk of developing both chronic respiratory diseases such as bronchitis and asthma as well as acute lung damage.

Many hoped that the availability of e-cigarettes would offer a safer alternative to traditional cigarettes and aid in smoking cessation, but research reveals that this has not been the case for kids and young adults. It turns out that kids who might not have started smoking are a lot more likely to start using e-cigarettes, thus resulting in a net increase of new nicotine users.

Marijuana

Adolescents' brains are more vulnerable to the adverse effects of marijuana than adult brains, especially when it comes to learning and memory. The human brain comes preinstalled with receptors for marijuana's active ingredient, delta-9-tetrahydrocannabinol (THC), not because it's our evolutionary destiny to be stoners, but because humans produce our own, naturally occurring endocannabinoids. These discoveries are recent, and research is still emerging, but it appears that our endogenous cannabinoids play a part in learning and help regulate anxiety.

The greatest concentrations of these receptors are in the hippocampus, so it's no surprise that memory impairment is a major side effect of pot use. Heck, it's no surprise to anyone who has ever been a pot smoker, known a pot smoker, or is aware of stereotypes about pot smokers. Cheech and Chong, Jeff Spicoli, Carl the

Gardener: all would have a difficult time negotiating rat mazes, the most common way to test memory formation and retention in rats.

Rat and human hippocampi are very similar in form and function, and studies have shown (rats!) that THC causes significant impairment when it comes to creating new memories, and another study (rats!) found that this impairment is far worse in adolescents than in adults. THC and its effects remain in the body for a long time, so long that people who smoke every few days are likely to have a persistent level of memory and problem-solving deficits and mental inflexibility. Heavy THC use is associated with increased gray matter volume in the hippocampus, possibly because pruning does not happen as it should, resulting in too many less than optimally efficient neural connections. These adverse effects in the brain can translate to adverse effects in life, such as academic failure, risky sexual behaviors, motor vehicle accidents, and social isolation, which can lead to more marijuana use, and so on and so on.

Marijuana use can also cause slower mental processing speeds (see Cheech and Chong et al.), unstable moods, decreased perseverance and lethargy, inflexible thinking, decreased attention, and motor impairment. Motor impairments may be due to the presence of cannabinoid receptors in the cerebellum and basal ganglia, both regions of the brain that regulate our movement. Marijuana can also increase heart rate while lowering our heart's efficiency, so it's a good thing that pot use and strenuous athletic endeavors don't tend to go together. In fact, the way many remember the difference between the two major strains of THC, indica and sativa, is that indica will sedate you right "in da couch."

Can marijuana cause psychotic breaks or schizophrenia? The complicated answer is maybe. For adolescents who are genetically predisposed to a specific abnormal gene expression that can lead to psychosis, the threat is real, and if I knew my kid had this abnormal gene expression I'd hammer his higher-risk status home to him on a fairly regular basis. However, for adolescents who do not have this genetic quirk, the risk is very, very low.

Legalization and the media frenzy around the epidemic of "hard" drugs such as opiates and methamphetamine might lead kids and parents to believe that pot is no big deal, and in cognitively mature adults, it may not be. Even given all the caveats around animal models and limited sample sizes, the research comes down on the side of delay, delay, delay. The younger kids are when they start using marijuana, the heavier their use is likely to be over their lifetime, and the more significant their cognitive decline and deficits are likely to be. Some of the damage done to the brain by marijuana use during adolescence is reversible, but a significant portion of it is not.

Opiates

When I talk about opiates, I am talking about both the prescription opiates teens may find in the medicine cabinet and illegal "street drugs" like heroin. They are, for all chemical and practical purposes, the same thing. More than one in five Americans will take some form of opiates during their lifetime, and those opiates live up to their hype as some of the most addictive (second to nicotine) and deadly drugs of abuse.

Opiates work because they act like our body's own natural analgesics (painkillers). When people (not me, mind you) talk about runner's high, it's an exercise-induced burst of these endogenous analgesics that they are banging on about. When accident victims talk about not feeling any pain despite having lost a limb, they also have the endogenous analgesics our bodies create in response to trauma to thank. Our endogenous analgesics block the transmission of pain and can either cause hyper-vivid consciousness that allows us to survive when being attacked, or envelop us in a warm, floating, dreamlike state. Opiate drugs are effective (and addictive) because they fill our naturally occurring opiate receptors with a more potent opiate than our bodies could ever produce on their own, in much greater quantities. Drugs like opium, heroin, codeine, fentanyl, oxycodone, and morphine (named for Morpheus, the Greek god of dreams) and their many variants make you feel, as one of my students put it, "like I'm being held by God." Even the memory caused him to mentally drift, and I quickly changed the subject to something less triggering. What he craved in that moment—and continues to crave if he's like all the other opiate users I know—is relief and return. Relief from the emotional, psychic, and physical pain of a childhood in abusive group homes, and a return to a high so appealing that author and former opiate user Tracey Helton Mitchell describes it in her memoir *The Big Fix* as "the best orgasm, the best hug, and the warmest blanket all wrapped up into a pile of *ahhh yes!*" No wonder Dorothy wanted to sleep forever in her vast field of poppies.

Opiates are dangerous because while we have lots of different opiate receptors for different kinds of neurotransmitters, opiate drugs activate every one, all at once. Opiate receptors are highly concentrated in the pons and medulla, areas of the lower brain that regulate breathing, and since they are often taken in combination with other drugs, their effects can be extremely dangerous. Opioid overdose can happen with the first dose or after twenty years of using because while tolerance for the high may increase, no such tolerance occurs in the breathing centers of the brain. Some other side effects of opiate use include drowsiness (nodding off, or "going on the nod"), confusion, dizziness, flushing in the skin, itching, vomiting, nausea, impotence, suppressed immune function, constipation (opiates halt digestion), and heart attacks.

Regular opiate users can have difficulty learning, problem-solving, and making decisions. Some researchers believe that repeated hypoxia, or lack of oxygen, that happens during opiate highs may lead to brain damage and loss of cognitive function in chronic users. Long-term studies of teens who start taking opiates at an early age reveal they are at higher risk of academic failure and dropout, diseases like hepatitis C and HIV (which can be transmitted by needles), and criminal behavior.

Stimulants

Stimulants do exactly what their name implies: they stimulate the body and brain, decrease fatigue, and can increase attention while producing feelings of euphoria and invincibility. This euphoria comes from a deluge of dopamine and a trick the drug uses to keep the floodgates open. Author, neuroscientist, and former methamphetamine user Marc Lewis explains: "This excitement and potency, this sense of efficacy and joyful enthusiasm, result from large amounts of dopamine, pooling in the synapses of my ventral striatum . . . it releases dopamine from its storage sites by tricking a clean-up molecule to work in reverse, emptying dopamine back into the synapses, rather than back into storage, after it has done its work."

With continued use, stimulant users may engage in obsessive, repetitive actions such as skin picking, pacing, and manic fits of "work" that look and feel productive to the hepped-up user but often result in output that's flawed or nonsensical. Heavy users can experience psychotic breaks that look a lot like schizophrenia and often require hospitalization. Chronic methamphetamine use can change the brain permanently, in ways that look a lot like premature aging. The effect of this is cruel: long-term users systematically reduce their ability to feel pleasure by reducing the amount of dopamine and serotonin that's available in the brain. Methamphetamine highs, in other words, rob kids of their ability to experience natural highs or even normal levels of happiness.

Sedatives

There are a few different types of sedatives, and when used by physicians, they are extremely effective in reducing anxiety, promoting sleep, and calming seizures. When used without a physician's guidance, sedatives cause memory impairment, extreme fatigue, and dangerous levels of sedation. On their own, sedatives can suppress breathing and heart rate to the point of death (except for benzodiazepines, which just make you sleepy and give you amnesia) and alter consciousness and response times enough to turn a car or other vehicle into a deadly weapon.

Sedatives are especially deadly when taken in combinations with other drugs, like opiates and alcohol. Not in rare cases deadly, but routinely deadly. I'm not one for fearmongering, but this is where I make an exception. Ask anyone who treats substance abuse and they will tell you, yes, opiates are dangerous, but it's the combination of opiates and sedatives that kills people.

There's been a lot of press lately about Ambien causing people to do bizarre things in a strange twilight sleep, including walking, talking, eating, sex, shopping, emailing, and driving, all lost to the memory impairment Ambien creates (again, by preventing the brain from creating memories in the first place). This effect is so common that a popular web comic character, Ambien Walrus (catchphrase: "Come with me on an adventure you'll never remember"), was created to personify the many hyphenated effects of taking Ambien. Sleep-order an entire ham, he coaxes. Sleep-order lobsters to be shipped overnight to your ex-girlfriend, he urges. For the record, these are all real chronicles with Ambien Walrus posted to Reddit's Ambien sub forum. These accounts are funny until you read the news stories about Julie Ann Bronson, who took Ambien, went to bed early, and woke up in a jail cell to the news that she'd run over three people with her car.

Speaking of news stories, one need not look much further than the death of Michael Jackson for evidence that even under physician supervision, sedatives can be lethal. What started for Jackson as a way to ease anxiety and get to sleep at night ended up as a sedative (propofol) and benzodiazepine overdose administered by his Mayo Clinic–trained, licensed, and board-certified personal physician, someone who understood complex dosing schedules and drug interactions and still managed to kill his patient.

Hallucinogens

These drugs can be divided into four loosely associated groups based on their psychedelic effects. First, we have the traditional psychedelics, drugs that cause psychological "trips" like LSD, mescaline, psilocybin, and ayahuasca. While it's true that the main negative side effect of the psychological hallucinogens such as LSD is a bad trip, they can also cause nausea, increased or decreased heart rate, chills, anxiety, and facial numbness. Injury from accidents caused by misunderstanding or misinterpreting the user's immediate environment can happen. Yes, flashbacks are a real thing, but usually in heavy users. Psychotic breaks, a well-publicized but much less likely side effect of hallucinogenic trips, can occur in 1 to 3 percent of instances of use.

Next we have the so-called horse tranquilizers or dissociative anesthetics like PCP and ketamine. Thankfully, PCP or "angel dust," the stuff of my middle school drug-scare nightmares, is less popular these days. It's really horrible stuff. It absorbs into the body quickly and lasts a long time (four to six hours for peak amounts, but it remains in the body for twenty-four to forty-eight hours). High doses of PCP are lethal and can cause psychosis with regular use. Combine PCP or ketamine with alcohol or sedatives and once again, we are in death via respiratory failure territory. Ketamine can raise body temperature, as can ecstasy, so combining these two drugs is especially dangerous.

The belladonna alkaloids like atropine, scopolamine, and jimsonweed tea should not be messed with. To get a hallucinogenic effect, you'd have to consume a dose that hovers around lethal. Finally, there is dextromethorphan, the active ingredient in cough syrup. In order to get the hallucinogenic effects, you'd have to exceed the recommended dose (as in, the whole bottle), so not only do you have to worry about the side effects of dextromethorphan itself (high blood pressure, nausea, vomiting, high body temperature, confusion), some over-the-counter cough syrups include other drugs such as acetaminophen, which is highly toxic to the liver.

Ecstasy

Ecstasy is a synthetic drug in its own category, an entactogen or empathogen, used most often in clubs or at raves to, as the name implies, create a sense of empathy and closeness to other people while lowering inhibitions, anxiety, and fear. Studies in people (and rats) show it causes both an amphetamine-like high and stimulant response in the body as well as a mild hallucinogenic effect in the brain, paired with teeth clenching, dry mouth, nausea, blurred vision, chills, sweating, muscle cramps, increased heart rate and blood pressure, and, at higher doses, seizures. While the details of how this drug works in the brain are still unclear, as are studies on long-term damage to the brain, the pleasure of an ecstasy high is caused by a massive, unnatural flood of serotonin in the brain. As the body works overtime to balance out the serotonin blitz, the good feelings are followed by a rebound, or neurochemical overcorrection, plunging the user into depression, anxiety, and exhaustion. This rebound effect increases with repeated or chronic use, and long-term damage to serotonin receptors is possible in some users. When paired with the heat, excitement, and activity of a club, dehydration and overheating are common side effects of these drugs. It's often referred to as a "love drug," but the irony is that ecstasy-induced love can't easily be consummated, as orgasm can be difficult to achieve on the drug.

Inhalants

Kids inhale lots of chemicals to create a buzz or high, from the gasoline in your car to the duster spray in your office and the solvents in your basement. These inhalants generally create a brief euphoria, stimulation, dizziness, and loss of inhibition or consciousness followed rapidly by depression that may or may not be accompanied by hallucinations. In higher doses, solvent intoxication can look a lot like alcohol intoxication, including slurred speech, incoordination, and vomiting. The rates of inhalant use continue to drop, but it's important to remember that it's mainly really young kids who use them because they can be found anywhere. Adverse effects in the short term include airway damage (from rapidly expanding compressed gas, such as dusters), burns (these chemicals are often highly flammable), and loss of oxygen while the chemicals are being inhaled instead. Damage to the cerebellum has been observed in users of toluene (a solvent), which can interfere with motor coordination and learning. Long-term use harms other systems as well, causing heart, lung, kidney, and liver damage.

This is not a comprehensive list of all drugs kids take, nor is intended to be. You can find information on the most commonly used drugs on the websites for the National Institute of Drug Abuse, drugabuse.gov, or the Partnership for Drug-Free Kids, drugfree.org.

CHAPTER 4

NOT MY KID
Who Gets Addicted, and Why

I love watching my students move from one class to the other, when the relative order of structured academic time devolves into unfettered teenage chaos. It used to make me nervous, this clamor of teenage posturing. But after twenty years of teaching teenagers, I have realized these unguarded intervals are an incredible opportunity to see who my students really are, in all their mercurial, unmannerly selves.

Kevin enters first, as usual. He chooses a seat up front and tosses a tattered copy of Jon Krakauer's *Into the Wild* onto the table with a flourish.

"Done?" I ask.

"Done," he says and smiles. "What's next?"

"What do you want? More nonfiction? Maybe take a one-eighty into some fiction?" I ask as I walk over to the independent reading shelf.

I offer a few suggestions and he stands in front of the book-shelf, considering his options. In his khakis, T-shirt, and Red Sox cap, he's the picture of New England boarding school chic. I tap my head then point to the cap, and he quickly removes it. I smile and offer a thumbs-up in return. He grabs two or three of my rec-ommendations and heads back to his desk.

On the way there, he accidentally bumps into Alex, who has planted himself in the middle of the aisle, where he's animatedly telling a story to a crowd of boys. Alex pauses his story for a mo-ment and stares at Kevin's back, deciding whether or not to turn the bump into a confrontation. The back of his neck begins to redden, and I hold my breath. Fortunately, the desire to finish his story wins out over his impulse to fight with Kevin, and the first crisis of the day is averted.

Katie and Destiny walk into the classroom together and con-spicuously ignore Alex and his entourage. Katie hates him and Destiny is crushing on him, but as the only two girls in the class, they have decided to maintain a united front where the boys are concerned.

There's a new kid in class this week, and he wanders in last, hood up and eyes down. He opts for a desk in the very back of the room and immediately puts his head on the table as if he plans to sleep. That's fine, for now. This is his first week here, in our little English classroom in Vermont, and he's still pretty pissed off. Usu-ally, I would introduce myself, but I don't have to. We already know each other. David was my student the year before, the first time he got kicked out of high school and packed off to rehab.

I don't ever like to bet against my students, but when David was discharged last time around, I knew I'd be seeing him again. He left early, having convinced his parents that he did not need to be in rehab.

"This place is bullshit. I don't belong here, with *these addicts*," he said.

These addicts are my students. They are not necessarily representative of the national drug- and alcohol-using adolescent population, mainly because New Hampshire and Vermont are uncommonly White, and because kids in rural rehabs take more prescription painkillers and begin using at an earlier age than their urban counterparts, but they are a pretty average group of New England kids. One thing I have learned over the past five years with my students is that no parent is ever in a position to say, "Not *my* kid."

Addiction exists in all ethnicities, socioeconomic groups, and geographic regions of the United States, and kids consume over 10 percent of the alcohol sold in this country. In any given month, between 8 percent (eighth graders) and 33 percent (twelfth graders) of American middle and high school students drink some alcohol, 10 percent take some illegal drug, 18 percent drink enough to count as a binge, 8 percent drive after drinking, and 20 percent have ridden with another person who has been drinking. Teens are the biggest abusers of prescription pain pills, stimulants, and anti-anxiety drugs, and consequently, drug overdoses in young adults have increased more than fourfold over the past decade.

Once we, as parents, teachers, coaches, and pastors, accept the scope of adolescent addiction, we can resist the powerful lure to deny its reach into our homes, schools, and communities. It's comforting to believe that our kids are strong enough, safe enough, smart enough, or supported enough to just say no. But it's only when we can admit, "Yes, maybe even my kid," that we can begin to see our own children with clear eyes and learn to recognize the risk factors that cause them to get tangled up in this dangerous behavior. In her memoir of her daughter's opioid addiction, *If You*

Love Me, author Maureen Cavanagh writes, "[T]he saddest thing I hear so many young people say is 'I didn't come from this.' I hear it again and again, 'I didn't come from this.' Their hands are up in disbelief. They are referring to their clothes, their teeth (or lack of them), or their surroundings. No sheets, possessions in garbage bags. 'I didn't come from this.'"

No matter where they come from, children share a predictable list of developmental stages, life experiences, and traits that lead them to seek out, experiment with, and become dependent on drugs and alcohol. When researchers ask kids why they drink or take drugs, many report they do it to feel better, to relieve emotional and physical pain, or to concentrate, relax, decrease anxiety, sleep, and cope with their problems. Of course, some say they like to experiment and that they like the way drugs and alcohol make them feel, but the vast majority of addicted kids are caught in a powerful, self-perpetuating cycle of self-medicating their pain, anxiety, and trauma. A kid like David, for example, self-medicates his untreated hyperactivity disorder and the pain of being abandoned by his father with a combination of pot, opioids, and alcohol. Given the choice between living in his chaotic brain, awash in his emotional pain, or numbing himself into oblivion, David chooses oblivion.

David's untreated hyperactivity disorder, his abandonment, academic failure, and unstable home life are what the National Institute on Drug Abuse (NIDA) refers to as "risk factors" for addiction. Addiction gains power over children in the presence of risk factors and loses power when exposed to "protective factors," but it's important to remember that lots of kids grow up under the weight of multiple risk factors, and never develop substance dependence or abuse. A risk factor for one person may not be a risk factor for another. However, experts agree on one thing: the sooner parents and teachers begin arming kids with these protective factors, the more effective they will be in preventing addiction.

I think of kids' substance abuse risk as existing on a justice scale, the old-timey kind with two pans suspended on either end of a beam. One side holds the risk factors for substance abuse, the other holds protective ones. Our goal is to make that protection side as heavy as possible, to outweigh whatever risk factors weigh the other side down. Risk factors combine and compound over time and if one, such as early childhood aggression, begins to bleed into another, such as academic failure or social ostracism, they can become very difficult to untangle. The earlier we start heaping on the protective factors, the greater our chances of raising a kid who makes it to adulthood substance-free.

Genetics

According to psychiatrist and substance abuse researcher Mark Schuckit, "substance abuse disorders are complex, genetically influenced conditions where genes explain up to 60% of the picture." The other 40 percent, he concludes, is to be found in an individual's environment and epigenetics (the connection between environment and genetics). There's not much we can do about changing our genetics, save for subjecting any potential baby daddy or baby mama to thorough DNA analysis and a genealogy background check. There's a technology called CRISPR (Clustered Regularly Interspaced Short Palindromic Repeats) that is being used to alter the genes of simple organisms, which are then used to edit human genes. It's freaky, it's scary (particularly given that in 2018, a Chinese scientist claimed to have created two CRISPR-edited babies,

Lulu and Nana), but many agree that CRISPR technology has the potential to eliminate or greatly reduce genetic disease. Chemical dependence originates in part in our genes, but solutions may ultimately reside in our genes as well.

Epigenetics

Addiction is not passed on to new generations solely through DNA. Researchers in the field of epigenetics, the study of the chemical processes that influence how genes express themselves in the body, have found that genes can be turned off or turned up in response to our life experiences. An addictive family tree can be a twisty, gnarled thing, where nature and nurture can't be completely pruned apart. The nature of our genes and the nurture of our childhood experiences are so intertwined that even our parents' childhood traumas live on in our bodies.

The word "epigenetics" was coined in 1939 by geneticist Conrad Hal Waddington and means, literally, "above the genes." Its meaning has evolved over time, and today, scientists refer to epigenetics as the science of how our environment impacts our DNA. These environmental factors, including trauma, exercise, sleep deprivation, diet, mental illness, and stress, chemically alter DNA through the addition or deletion of chemical compounds such as methyl groups, which in turn change the way the gene is expressed, but they do not change the DNA itself. Recent research indicates that non-genomic traits may be heritable in humans (it's been documented in other, nonmammalian species), meaning that some nongenetic traits may pass from parents to child via exposure to stress or disease in utero or early in childhood. For example, children who were in utero during the "Dutch Hunger Winter" of

1944–45 had higher rates of obesity, cholesterolemia, diabetes, and schizophrenia, and died at higher rates than the generation before or after. The theory is that the famine, like other childhood traumas, altered the epigenetics of the children born during that period. After further examination, researchers found methyl group changes on genes related to higher body mass index. As science writer Carl Zimmer explains, "Perhaps the Dutch Hunger Winter added a methyl group to fetuses born to starving mothers, which made the PIM3 gene [a gene involved in metabolism] less active—and continued to do so for life." As the science of epigenetics evolves, the line between nature and nurture gets fuzzier and fuzzier.

Addiction at Home

In the words of just about every addiction expert, "Addiction is a family disease," both because of the genetic and epigenetic footprint of the disease and its tendency to negatively impact the lives of everyone in the family. Children who grow up with addicted parents are "primed" genetically, emotionally, and experientially for addiction. They are not only more likely to become alcoholics themselves, they are also more likely to marry an alcoholic even if they are not alcoholics themselves, thus perpetuating the intergenerational cycle of addiction.

Part of the reason this happens is that these kids are born into a lifestyle built on what researchers call "experiential factors." Children of substance-abusing parents learn how to adapt to lives of chaos, and most get really good at it. Some get so good at it that they come to expect the chaos, and intentionally or unintentionally invite it into their lives. These experiential factors are, in effect, an education in managing addiction: lessons on how to care for a future addicted spouse, on the painful burden of shame, on how to keep a family secret and a smile on your face, pretending that "everything's fine" while life at home is anything but. Addicted parents often can't parent as effectively as sober parents, and this puts their children at risk for physical harm, neglect, depression, hopelessness, and, once again, addiction.

Adverse Childhood Experiences

Our understanding of addiction has evolved tremendously over the years. What was once viewed as a failure of character or a lack of self-control is now understood as a disease or, more recently, as a developmental disorder. Daniel Sumrok, director of the Center for Addiction Science at the University of Tennessee, believes we should not even call drug- and alcohol-seeking behavior "addiction." We should be calling it "ritualized compulsive comfort-seeking," an expected, predictable response to painful life experiences. Taking drugs or drinking alcohol in order to alleviate the pain of early life, he believes, is as understandable as taking aspirin for a headache. My rehab students take drugs, he would likely say, because they are self-soothing, blunting their physical and emotional pain the only way they know how. Our goal, then, should be to identify the reasons kids need to self-medicate, and help them lessen that pain before they feel the need to resort to drugs and alcohol for relief.

Childhood experiences, both negative and positive, have an enormous impact on health and welfare. When a person endures what the Centers for Disease Control and Prevention (CDC) calls an "Adverse Childhood Experience," or ACE, they are far more likely to have chronic health conditions, and to engage in high-risk behaviors that threaten their health and safety. In 1995, just as physicians were beginning to understand the impact of these adverse experiences on lifelong health, the CDC and the Oakland-based health care consortium Kaiser Permanente set out to quantify the degree to which these experiences shape our health and wellness. To that end, Vincent Felitti, chief of Kaiser Permanente's obesity clinic, and Robert Anda, a medical epidemiologist with the CDC, surveyed more than 17,000 patients about their current health, lifestyle, and behaviors. The resulting report on these patients defined ten types of abuse and household dysfunction that have the greatest impact on lifelong physical and mental health, including the likelihood of becoming addicted to drugs and alcohol. Felitti sums up the results of the study: "Adverse childhood experiences are the main determinant of the health and social well-being of the nation."

Most of us hear the phrase "adverse childhood experience" and immediately think of kids raised in poverty by drug-addicted parents. I certainly did. Growing up in a White, wealthy neighborhood where I felt insulated from trauma, I allowed television and movies and my own implicit bias about race and wealth and power to shape my assumptions about childhood trauma and its victims. Poor kids get beaten. Black and Brown kids have drug-addicted parents and live in violent neighborhoods. Uneducated parents neglect their children.

These beliefs are not just wrong, but they also feed our denial and keep us from facing the reality that adverse childhood experiences happen in every demographic, every geographical region, and to every race and ethnic group. Nadine Burke-Harris, pediatrician and author of *The Deepest Well: Healing the Long-Term*

Effects of Childhood Adversity, notes that the ACEs study upended the myth that poverty is at the root of poor health outcomes. ACEs, and the toxic stress they create, she writes, are "astonishingly common—67 percent of the population had at least one category of ACE and 12.6 percent had *four or more* [emphasis hers] categories of ACEs." What's more, the subjects of the ACEs study were "Solidly middle-class San Diego [. . .] 70 percent Caucasian and 70 percent college-educated."

Adverse experiences and the toxic stress they create is happening everywhere. If we can accept this reality, we can get serious about prevention. If not, we are destined to perpetuate the conditions that feed it. Felitti claims that the billions of dollars we have spent on drug and alcohol prevention have been spent in all the wrong places. "Our findings are disturbing to some because they imply that the basic causes of addiction lie within *us* [emphasis his] and the way we treat each other, not in drug dealers or dangerous chemicals."

So take a deep breath and be prepared to view your family history with a clear eye, unobstructed by the blinders of shame and blame. Remember, ACEs are, in the words of one of the authors of the study, "surprisingly common" in every demographic, so most people you know have experienced one of them. The CDC sorts them into three categories: abuse (emotional, physical, and sexual), household challenges (witness to violence, substance abuse, mental illness, criminal behavior, and separation or divorce), and neglect (emotional or physical).

Adverse childhood experiences put people at increased risk for negative health and behavioral outcomes such as heart, liver, or lung disease, as well as a higher incidence of smoking, alcoholism, and drug use; early initiation of sexual activity, multiple sexual partners, teen pregnancy, and fetal death; poor academic achievement and subsequent financial stress, depression, and suicide attempts; and a higher likelihood of experiencing sexual violence and intimate-partner violence. People with an ACE score of four on the CDC's assessment are seven times more likely to develop alcoholism than people with a score of zero. Some ACEs, such as sexual abuse, have been identified as particularly powerful predictors for substance abuse.

"If we could eliminate all violence—bullying, sexual and physical abuse, sexual harassment—we could prevent sixty-six percent of binge drinking in twelve- to eighteen-year-olds. Sexual abuse accounts for twenty percent of binge drinking and sexual harassment for fifty percent," says Elizabeth Saewyc, director of the Stigma and Resilience Among Vulnerable Youth Centre at the University of British Columbia School of Nursing.

The CDC and Kaiser discovered a "graded-dose relationship" between ACEs and negative outcomes, meaning that the more adverse experiences a kid accumulates and the higher his ACE score, the higher the severity and intensity of the negative outcome. For example, "a male child with an ACE score of 6, when compared to a male child with an ACE score of 0, has a 46-fold (4,600%) increase in the likelihood of becoming an injection drug user sometime later in life."

In addition to the list of adverse childhood experiences with the greatest impact on lifelong health and wellness enumerated by the CDC, Burke-Harris has added other experiences to the questionnaire she uses in her pediatric practice that reflect the wide range of childhood experiences that can make kids sick, unhappy, and addicted:

- Community violence

- Homelessness

- Discrimination

- Foster care

- Bullying

- Repeated medical procedures or life-threatening illness

- Death of caregiver

- Loss of caregiver due to deportation or migration

- Verbal or physical violence from a romantic partner (teen questionnaire)

- Youth incarceration (teen questionnaire)

As Felitti noted in a paper written in 2004, after the first wave of research had been completed, "an initial design flaw was not scoring subtle issues like low-level neglect and lack of interest in a child who is otherwise the recipient of adequate physical care." Other experts on addiction, most notably addiction counselor and interventionist Candy Finnigan, point to adoption as a significant factor in the development of addiction due to persistent feelings of abandonment. So the kinds of adversities that enter into the equation can be obvious or subtle but they all have to be taken into account.

I have been teaching for twenty years and working with kids as a juvenile attorney, coach, and community volunteer for even longer than that, and I don't know many parents who are in a position to say "Not my kid." These adverse experiences affect all of us, and it's going to take all of us to protect our children from the harm they inflict.

Toxic Stress

Kids are incredibly stressed out. I know this both through my own experience talking to thousands of kids a year during school visits and in my own classroom, and through research on the state of the American adolescent. According to the "Stress in America" report published by the American Psychological Association, the most stressed-out people in this country are adolescents between the ages of 15 and 21. "The Kids are Not Alright," a 2010 paper released by the American Psychiatric Association, details the stress affecting kids, and indicates that it's causing more than emotional distress; it's causing physical symptoms like headaches and sleep-lessness, symptoms that parents are failing to recognize as signs of anxiety. More than 30 percent of kids report headaches within the past thirty days, and 44 percent of children report difficulty sleeping. The Boys Town National Hotline reviewed over 830,000 calls received since 2012, and they report a 12 percent increase in complaints around anxiety, depression, and suicidal ideation over the past five years.

Worry about substance abuse is part of adolescent stress. Fifty percent of surveyed adolescents report "that at least one person they know has been told they are addicted to or have a problem with drugs and alcohol" and 39 percent cite the opioid epidemic as a source of stress. Twenty-five percent say they would not know where to find help if they had a problem themselves, and 39 percent say they would not know where to get help if a family member or friend was struggling with substance abuse. Stress can cause serious physical symptoms in the short term, but it also raises risk for all sorts of physical and psychological conditions over the long term. Anxious, stressed-out kids are at increased risk for addiction because drugs and alcohol seemingly can make problems go away, at least until the buzz wears off. Kids self-medicate with alcohol and drugs (as well as with food, sex, the internet, self-injury, or lots of other forms of emotional release) in order to escape, even for just a moment, the discomfort of long-term threats, harm, or challenges. If you were to ask teens to rank their stress level on a scale of 1 to 10, the kids with stress levels of 6 or higher are three times more likely to have used marijuana than kids with stress levels of 5 or lower, and twice as likely to have used alcohol or tobacco.

Because ACEs are long-term, prolonged, negative experiences, they tend to cause so-called toxic stress. Toxic stress is particularly harmful because it doesn't just cause short-term symptoms like headaches and sleeplessness; it actually changes the way our bodies cope. It alters our stress-response systems and brain architecture, and interrupts learning. In Burke-Harris's practice, for instance, her patients with four or more ACEs were 32.6 times more likely to be diagnosed with learning and behavioral problems.

Unfortunately, toxic stress is not a diagnosis in and of itself and is unlikely to be isolated as a cause of physical, emotional, or cognitive issues by even the most observant physician. Even Burke-Harris, an expert in the effects of ACEs and the toxic stress they engender, admits that she spent years missing the connection between her patients' experiences and their physical complaints, psychological conditions, and learning disabilities. It wasn't until the results of the ACEs study were released and she articulated her own additional list of adverse experiences that she was able to start helping her patients heal and move on from their trauma.

Academic Failure

According to the National Institute on Drug Abuse, children who struggle in school when they are between 7 and 9 are more likely to be using addictive substances when they reach 14 or 15. Academic failure is yet another risk factor that can emerge either as a primary risk or as a consequence of some other risk factor such as aggressive behavior or adverse experiences at home. No matter the cause, it's imperative that parents, schools, and learning specialists work together to figure out what's going on and confront the issue. So many factors can cause academic failure, from learning disabilities to behavioral issues, and it's important to find the underlying cause of the deficit.

Untreated attention deficit hyperactivity disorder (ADHD), in particular, has been identified as a major risk factor for addiction. My rehab classroom always contained at least one or two students who struggled with ADHD, and they expressed incredible frustration over their inability to pay attention in school. In his book *In the Realm of Hungry Ghosts: Close Encounters with Addiction*, Canadian physician and childhood trauma expert Gabor Maté writes,

[A]s many as 35 percent of cocaine users who presented for treatment met the diagnostic criteria for childhood ADHD. In another study, as many as 40 percent of adult alcoholics were found to have underlying ADHD. People with ADHD are twice as likely as others to fall into substance abuse and nearly four times as likely as others to move from alcohol to other psychoactive drugs. People with ADHD are also more likely to smoke, to gamble, and to have any number of other addictive behaviors. Among crystal meth addicts a significant minority, 30 percent or more, also have lifelong ADHD.

Academic failure can happen for many reasons, and of course it's important to get answers as to why. However, while we are investigating the why, while we line up the testing and counseling and search for ways to support their learning, the most important thing we can do as parents and educators is give kids reasons to believe their lives will get better. Kids who struggle to keep their heads above water at school often feel worthless, helpless, and hopeless. When kids feel hopeless, they are vulnerable to increased rates of substance use, absenteeism, expulsion, depression, inactivity, and sleeplessness.

On the other hand, kids who believe their future will be better than their present do better across the board. Hope increases academic performance, graduation rates, career success, and happiness and provides a psychological buffer from the effects of negative life events. One study of law students found, "A student's level of hope predicted his or her law school ranking better than the LSAT (the law school entrance exam)." In the words of the late Shane Lopez, psychologist and hope researcher, "[H]ope predicted test scores and term GPA when controlling for previous grades, intelligence, and other psychological variables (like engagement, optimism, and self-efficacy)." When Lopez and other researchers at Gallup asked one million people whether they had smiled or laughed the day before, the hopeful said yes much more often than the hopeless.

The greatest challenge of teaching in my rehab classroom was the palpable lack of hope. That void stole my students' laughter and joy and rendered so many of my go-to tips and tricks for boosting morale and performance worthless. The good news is that kids are wondrously resilient. For most of the kids in my classroom, hope wasn't absent, it was merely buried under years of evidence to the contrary: the consequences of bad decisions, their low expectations, and their self-defeating behaviors. One of the most important parts of my job, then, was to help them unearth the ruins of their hope and show them how to use it to build a better future, shore up their goals, and restore their laughter.

Early Aggressive Behavior

As mentioned earlier, when childhood risk factors are left untreated, they can combine and conflate to create other problems, leading to a chicken-and-egg puzzle of causality. Aggressive behavior is a prime example of a risk factor that can emerge as a novel behavior, but it can also be a secondary manifestation of other risk factors left unmanaged. Academic failure, for example, can morph into aggressive behavior as frustration and stress build up in a child.

No matter the cause, aggressive behavior needs to be addressed right away. According to the National Institute on Drug Abuse, children who exhibit early aggressive behaviors are more likely to be rejected and ostracized by their peers, get punished more often by their teachers, and (back to the whole chicken-and-egg problem) experience academic failure. This is why early intervention, support, and treatment is so important.

In some children, abnormally aggressive behavior can be identified as early as infancy and should be treated that early, too. While it's found less often in girls, aggression toward peers is one of the main reasons boys experience social ostracism and rejection, which is in turn yet another risk factor for alcohol and drug abuse.

Transitions

Kids are most vulnerable to substance abuse—especially initiating that first sip or drag or huff—during periods of transition. These transitions can be around stages of maturation, such as from childhood to adolescence or from adolescence to early adulthood, or they could be due to life changes, such as divorce, death of a loved one, or moving from one residence to another. The stress that arises from a transition is often linked to feelings of helplessness and lack of control, and transitions such as moving or starting a new school can make kids feel untethered and therefore especially sensitive to peer pressure.

Summer

Even summer, for all its fun and free play, presents a heightened risk of substance use. According to the 2019 National Survey on Drug Use and Health, "Thirty percent of marijuana use, 28 percent of cocaine use, 34 percent of LSD use, and 30 percent of ecstasy use was initiated in summer compared with other seasons." The authors conclude that the increase in drug use, and especially first-time use, may be chalked up to increased idle time combined with more opportunities to attend parties and music festivals during the summer months. As a parent, I love this data because it's specific. If I know that first-time use is more likely to happen in the summer, I will be sure to spend more time talking about these four substances in the late spring, and to discuss the specific risks of these four drugs. I'd also be more apt to talk about their exit plans in case of heightened peer pressure or supervision issues, and I'd even review our family contract regarding substance use and our agreement that we will always be available for a ride, no questions asked. The more specific the data on risk, the more specific and targeted my prevention efforts can be.

I've been teaching young addicts for years now, and even I am overwhelmed and frightened by the weight of all the risk factors I've just articulated. I am their writing teacher, so I get to read their most honest stories, reflections on abuse, divorce, violence, and intergenerational addiction. The points rack up as I read: ACE upon ACE upon ACE. I often wonder how any of them make it to adulthood intact, and yet, they do. Given enough support from the adults in their lives, lots of protective factors to inoculate them from all that risk, a belief in their own competence, and plenty of hope, even kids like David can lead happy, fulfilling, healthy lives free from substance abuse.

CHAPTER 5

TIPPING THE SCALES OF ADDICTION
The Protective Factors That Outweigh Risk

In the years I was drinking, my main focus was on protecting my right to drink, and keeping it a secret from my boys. From the day I got sober in 2013, however, my focus shifted to protecting my kids from the genetic and environmental risks I'd strewn in their path. Until very recently, I felt great about my efforts. My husband and I do our best to control the risks we can. I got and stayed sober; our kids go to annual well-child checks with a pediatrician they adore; they have good relationships with teachers and school administrators; and they have had access to mental health intervention when it's been needed. We have also been fortunate when it comes to the factors that are more difficult to control. We have never had to struggle to put food on the table or heat our home; we have remained married; our kids have solid, long-term friendships with kids whose families I adore and respect; and we have lived in a small New England village since the time our boys started school.

Then, in 2018, we left.

My husband had been unhappy at work for a long time, so when he was offered his dream job, I swallowed hard, slapped a smile on my face, and told him to go for it. Besides, I said, it's only two hours away.

Our older son, Ben, was cool with the news, as he was off at college and loved the area where we would be living. His one hesitation was that our move would put us within surprise-visit range of his college. Once we promised to call first, he was on board.

Finn, who was fourteen at the time, was devastated.

"You are ruining my life," he said, when we told him about the move. There was no yelling, no wild gesticulations, just a calm statement of fact, which was much, much worse.

I used the "it's only two hours" reassurance, but for a kid who could not yet drive, we may as well have been moving to Mongolia. His two best friends were within walking distance of his childhood home and the woods, fields, and ponds of our rural New Hampshire neighborhood were his comfort and his refuge. Two hours or twenty, away is simply away.

From a parenting standpoint, we could not have done much worse. Research reveals a steady increase in the numbers of kids who become addicted between the ages of 12 and 18, and their first use typically happens in seventh or eighth grade. We had voluntarily signed our adolescent boy up for an up-close and personal tour of the risk factors for adolescent substance abuse during his most vulnerable period of development on top of the genetic risk he already faced. A stressful physical and emotional transition? Check. Relocate to a state with permissive attitudes and laws about marijuana? Check. Sever ties with a peer group we trust? Check. Replace those peers and their supportive, loving parents with families we have never met? Check.

Before we moved, Finn had protections heaped on his metaphorical substance abuse scale: our wonderful, supportive village
of adults who love him; physical, financial, and emotional stability;
lack of stress; and his friends' parents looking out for him and providing healthy models for sobriety, support, and coping. Now that
I had thrown his balance off, my job was to figure out what I could
do to outweigh the weight of his risk with as much protection as
possible. Just as I can't control every risk factor, I can't implement
every single protection that might be available. My hope has to be
to keep the protection side heavier than the risk side, even if just
by a little bit. That's all parents can hope to do if we want to keep
our kids as safe as possible from substance abuse.

My goal in this chapter is to help adults understand what aspects of substance abuse we can control for our kids, but there are
so many we cannot. The National Institute on Drug Abuse (NIDA)
categorizes risk under domains: individual, family, peer, school, and
community. While we might be able to effect change on the level
of the individual, family, peers, and possibly schools, the community factors, such as the availability of drugs in a given city or town,
poverty, and lack of access to health care, are virtually impossible for
any one person to change on his or her own. Likewise, many of the
ACEs listed in chapter 4, such as poverty and community violence,
are systemic problems requiring systemic solutions.

Finally, I can't talk about these protections without acknowl-
edging that I am in a position of great privilege when it comes to
what I can control, in that I am White, relatively wealthy, have access to health care, was raised to believe in my own power to effect
change, and have a supportive family around me. If I were Black
or otherwise minority, poor, a single parent, indigent, or lacking in
family support, it would be more difficult for me to keep the risks
to my children at bay while heaping on protections.

Substance Abuse in the Home

As I've discussed, substance abuse is passed down to children through a combination of genetics, epigenetics, and environment. While we can't do much about the genetic inheritance, here are some ways to manage your child's epigenetic and family risk for substance abuse. If you are having trouble coming to terms with the stressors your family faces, feel free to start by mentally substituting the scary labels of "addiction" or "substance abuse," or "trauma" with "stress," because in the end, that's really what we are talking about, whether it's due to the Dutch Hunger Winter or a parent who drinks too much.

- CALL IN REINFORCEMENTS. When one family member suffers from substance abuse, everyone in the family needs emotional and mental health support, even kids who appear to be coping well. Children raised around substance abuse are often high-functioning people pleasers and perfectionists, and can suffer in silence for years before anyone thinks to offer mental health support. As with all risk and prevention measures, the earlier children get counseling to help them cope with the stress of an addicted parent or sibling, the better.

- BANISH SECRETS AND SHAME. Most families have secrets. In fact, research shows that most families harbor two to three of them. Family secrets, usually in the form of an unspoken agreement to ignore the elephant in the living room, are destructive, because secrets tear families apart and make room for more substance use. Growing up with alcoholism was painful, but not as painful as being told that what I was seeing and hearing did not exist, and even if it did, I was not allowed to talk about it. I adapted by being as perfect and low-maintenance as possible, behaviors that worked in the short term but set me up for a lifetime battle with anxiety and, of course, my own substance abuse.

- TALK OPENLY ABOUT YOUR FAMILY RISK. When children are aware of their genetic predisposition for substance abuse, that knowledge can serve as a positive, preventive factor. For example, my older son told me that he used his heightened genetic risk as an easy, readily available excuse not to drink during high school.

- KEEP ADDICTIVE SUBSTANCES UNDER LOCK AND KEY. Alcohol, as well as narcotics and other prescriptions, should be kept away from kids. One-third of teens see no problem with using drugs not prescribed for them and one in four teens has misused a prescription drug. Half of the teens who admit to misusing prescription drugs report they acquired those drugs from their parents' medicine cabinet. If you must have opioids or other narcotics at home, purchase a storage safe and keep it locked.

- GET MOVING. Physical activity has been associated with a lower prevalence of cigarette use and some harder drugs. Besides, exercise is not just good for the body, it relieves stress and benefits the brain. Learning new sports can also satisfy teens' cravings for novelty and stimulation, which in turn stimulate the growth of neurons in the brain. These new neurons can mean greater potential for learning, greater integration of higher brain and lower brain, and increased ability to self-regulate the natural impulsivity of adolescence.

- GET A PET. Human-animal interactions have been shown to improve behavior, mood, and interpersonal interactions; reduce symptoms of stress such as raised cortisol levels, high blood pressure, and elevated heart rate; reduce negative behaviors such as aggression; enhance trust levels and empathy; and enhance learning. Many of these effects are due to oxytocin, a hormone produced in the hypothalamus and released into the blood when we experience stimulus such as petting an animal, breastfeeding a baby, cuddling, sex, and the warmth of other trusted humans. Oxytocin, in turn, stimulates further social interaction by increasing eye contact, empathy, face memory, trust, and generosity. Oxytocin is also as close as an antidote to aggression as we are likely to find, as it's the hormone responsible for facilitating all forms of human bonding. The closer the human-animal bond, the more oxytocin is released, so stroking your own dog will produce a more calming effect than stroking one you've just met. As an added bonus, those animals are going to need exercise, which increases the chances that their owners will exercise more as well. Research shows the odds of getting enough exercise are between 57 and 77 percent higher among dog owners. If you can't adopt your own pet because of allergies or because it would add to the stress in your home, have your child or teen advertise their skills around the neighborhood as a dog walker for hire, or have them do what my father does: offer to walk the neighbor's dog for free, simply because it makes him happy to hang out with his buddy Bella once a day.

Build Kids' Self-Efficacy

In the face of adverse childhood experiences, one of the most powerful protections for kids is self-efficacy: their belief in themselves to succeed; to regulate their thoughts, emotions, and life; and to cope with challenges in a positive way. The theory of self-efficacy, as posited by psychologist Albert Bandura, is the foundation for so many other positive traits, including resilience, grit, fortitude, and perseverance. Self-efficacy is what gives kids a sense of control, agency, and hope, even when the world around them feels out of control.

A belief in one's own self-efficacy is, Bandura argues, "the foundation of human motivation, well-being, and accomplishments." When people possess a strong sense of self-efficacy, they are more likely to connect the future they want with the actions they must take to make that future happen. They will be more committed, diligent, and determined even in the face of obstacles and failures, and consequently, more likely to succeed (and be rewarded with a dose of dopamine for their efforts). If they fail, they will be more likely to pick themselves up and try again. If they prevail, their success reinforces and feeds feelings of self-efficacy and motivation, which cranks the positive feedback loop into high gear, fueling future success and confidence.

People with a strong sense of self-efficacy are more likely to be optimistic, motivated, confident, competent, adaptive, resilient, flexible, goal oriented, and self-driven. They are also more likely to set and achieve goals under their own steam, view obstacles as surmountable, have a lower fear of failure, and approach new challenges with the assumption that they can succeed. On the other hand, people with a weak sense of self-efficacy are pessimistic, inflexible, quick to give up, have low self-esteem, exhibit learned helplessness, get depressed, and feel fatalistic and hopeless. Not coincidentally, people who exhibit these traits are more likely to turn to drugs and alcohol to alleviate these negative feelings.

Lack of self-efficacy is a risk factor for substance abuse and other negative health outcomes, but when converted into its opposite and equal force, a strong sense of self-efficacy, it can be one of the most powerful protective factors we can give our children. In fact, Bandura found that a strong collective self-efficacy, such as in a family or a classroom, can mediate the effects of poverty and prior academic failure.

Self-efficacy can also serve as protection against peer pressure and promote open parent-child communication. Bandura asserts that perceived self-efficacy protects kids from peer pressure both directly and indirectly: first, if a teen believes he can resist peer pressure, he will be a lot more likely to do so, and second, he will be more likely to talk to his parents about episodes of peer pressure when they arise. On the other hand, according to Bandura, kids who don't feel as if they can resist peer pressure don't tend to talk to their parents about the things they do outside the home. And given that good parent-child communication serves as yet another protective factor, I'm sold. Sign me up. I'm going all in on self-efficacy.

The most effective, evidence-based substance abuse prevention programs have gone all in on self-efficacy as well. These programs tend to take place at school, but many of the same strategies work at home. In fact, school-based prevention programs work best when the lessons are echoed at home, reinforcing both the practical information and the endgame: reinforcing self-efficacy in children.

Here are some practical ways parents can boost kids' perceptions of their own self-efficacy and help kids with low self-efficacy get back on the right path:

- START WITH YOURSELF. Model, model, model self-efficacy for your kids. Start questioning your own assertions of "I can't" with "I can't *yet*," then turn that perspective outward, toward your children. The best response to our children's frustrated sigh of "I can't do it" is "You can't do it *yet*," a phrase that helps kids believe competence is not congenital, it is learned, and often hard won.

- BELIEVE IN YOUR CHILDREN AND MAKE SURE THEY KNOW IT. Kids who know their parents believe they are competent and have faith in them to make good decisions have higher levels of self-efficacy and self-confidence. When a kid feels trusted, they are more likely to act in ways that will perpetuate that trust.

- GIVE KIDS SKILLS. Praise alone won't give your child a sense of self-efficacy or competence; these things come from the actual experience of trying, doing, failing, trying again, and succeeding. When we tell our kids how wonderful, smart, and talented they are without giving them the opportunity to re-inforce our belief in them with skills, we risk raising, in Bandura's words, "self-confident fools." Give kids "Goldilocks" tasks, age-appropriate yet challenging tasks that help them stay engaged and challenged while granting opportunities to taste success. For a toddler, that may be as simple as learning how to fold a napkin or put the silverware in its proper place on the dinner table. Teach your older child how to make dinner from start to finish and see what they create on their own. Encourage your teen to take the family car to the garage and have that rattle behind the dash fixed. Recently, I planned for an extra hour at the airport on a family trip and handed control over to my teen as soon as we entered the airport terminal. For sixteen years, I treated him like little more than another carry-on bag, something to be pulled along behind me as I navigated ticket-ing, security, and unexpected delays, but now he knows how to navigate an airport on his own.

- SEE ONE, DO ONE, TEACH ONE. This is a common practice in medical school in which students watch a procedure, do a procedure on their own, then teach someone else how to do the procedure. Kids do not learn well from mere observation; they must have the freedom to muck about in the task once they see how it's done. Start early, well before kids head off to school, because, as psychologist Martin Seligman writes in *The Optimistic Child*, "Masterful *action* [emphasis his] is the crucible in which preschool optimism is forged," and optimism is an important stepping-stone to self-efficacy. My younger son is learning how to drive and no matter how many times I show him how I parallel park, he'll never really understand how to do it himself unless he spends some time in the driver's seat.

- SET INDIVIDUAL AND FAMILY GOALS WITH AN EYE TOWARD LEARNING, NOT ACCOLADES. Setting goals is essential in building what's called self-directed executive function, the skill of setting a goal and then executing the smaller tasks involved in achieving it. Self-directed executive function is where perseverance, fortitude, and competence are born, so let kids set their own goals based on their own concerns and interests, and support them as they work to achieve those goals. Start by setting your own goals and your plan of action. Yes, you have to talk to your kids even when you fail—especially when you fail. Success and accolades are great, but teaching kids to learn from their mistakes is how we help kids value learning and mastery, which is a part of building self-efficacy. In our family, we make it a point to articulate a few short-term goals each season, and one of those goals has to be a bit beyond our comfort level. We even write them down, and after a few months we pull those slips of paper out and check in with each other to see how those goals are progressing. If they are not going well, we talk about how to do better next time, and if they are going great, we celebrate our wins.

- BE OPTIMISTIC. Optimism is about more than seeing a glass as half full; it's a mindset that has a very real impact on physical and mental health. Optimistic children are better able to resist learned helplessness and depression, whereas pessimists are much more likely to give in to feelings of helplessness and are consequently at much higher risk of depression as well as physical illnesses that can result from the immune system suppression that depression can cause. According to Seligman, pessimistic kids see obstacles as permanent, pervasive, and their fault. Optimistic children, on the other hand, view setbacks as temporary, specific, and attributable to behaviors that can be changed. Parents can help kids move from a pessimistic to optimistic mindset by shaping the way kids view setbacks. For example, if your daughter says, "I got in trouble at school because I was bad and Mr. Martin hates me," guide her through these three steps. First, help her understand the difference between sometimes and always. Talk through her history with Mr. Martin so she can see that even if he is displeased with her today, his feelings about her are not permanent. Remind her of times Mr. Martin was supportive and kind, when he took interest in her learning or interests. Second, help her see that the cause of the setback is not pervasive, that if she changes her behavior (whatever caused Mr. Martin to become displeased with her), she can change or eliminate his displeasure. Finally, talk about the cause of the conflict with Mr. Martin. If she is to blame because of her own behavior, help her own that blame while helping her view her share in the blame as specific to a behavior rather than general to her identity. Instead of "I'm bad," help her understand "I did something bad." Instead of "I'm stupid," guide her toward "I did not understand what he wanted me to do in class today."

- MAKE FAILURES SPECIFIC, BUT GENERALIZE SUCCESS. As we shepherd children toward optimism by making setbacks as specific as possible, we also need to help them view their success as generally as possible. If your daughter has a good day with Mr. Martin, help her globalize that success. Instead of "I did well in math class because I paid attention," move toward "School is going well because I am doing all my assignments on time." Help her expand her success beyond the boundaries of one class or one day through her continued positive behaviors.

- BE CAREFUL WITH WHAT YOU SAY, BUT BE EVEN MORE CAREFUL IN WHAT YOU DO. Children hear and see so much of what we do, even when they appear to be ignoring us or willfully missing the point. Yes, I'm even talking about little kids with undeveloped language skills and teens who have perfected the art of ignoring parents. In 1994, when I worked in pediatric HIV research, I met a very young patient whose disease status had been kept from her because her parents were worried it would scare her and worsen her outcome. Imagine our surprise when we discovered she'd known all along. She confided in her nurse, "Please don't tell my parents I'm sick. I don't want them to worry." Young children and adolescents hear and see our family's truths, no matter how hard we work to keep them sheltered.

- BE SPECIFIC IN YOUR PRAISE. General praise, such as "Good job!" is worse than useless when it comes to bolstering self-efficacy in kids because it has no real meaning, and undermines the power of any praise that might help kids feel more competent and powerful in the world. Aim for behavior-specific praise that reinforces practices you want to encourage, such as, "I'm so proud of you for sticking with that project even when you got frustrated." Behavior-specific praise describes the desired behavior, is specific to the child, and offers a positive, clear, statement of praise.

- BE HONEST IN YOUR PRAISE. If your child is tone-deaf, don't tell her she's the next Beyoncé. Children are not stupid, and when we lie to them about their abilities, they know it, and begin to question our judgment and honesty. To quote psychologist Erik Erikson, "Children cannot be fooled by empty praise and condescending encouragement. Their identity gains real strength only from wholehearted and consistent recognition of real accomplishment." Be honest about kids' strengths and weaknesses while guiding them toward continued improvement, regardless of whether they have natural talent or have to work a little harder than their peers.

- DON'T GO OVERBOARD WITH YOUR PRAISE. Experts on the use of behavior-specific praise in the classroom recommend a 4:1 ratio of praise to reprimand, a ratio I have tried to maintain with my own students and children. I teach and parent older teens, but this guideline is effective for kids of any age. Research shows it not only boosts good behavior, but also creates a sense of community and positivity that helps kids hear our constructive criticism when it inevitably comes.

Health Care

Your child's pediatrician can be one of your greatest allies in preventing substance abuse. Many parents are aware that substance use is a health risk, but given that relatively few believe that *their child* is at risk, screening at annual well-child checks can be a powerful tool for bridging that perception gap. More than 90 percent of adolescents ages 12 to 17 have access to a place where they receive health care (either a physician's office or a clinic), and 83 percent of adolescents get an annual physical, so health care providers are in a great position to screen kids for high-risk behaviors and intervene before patterns and habits emerge. According to a survey designed to test attitudes about their willingness to talk to health care providers about risky health behaviors, most teens said they trust their health care providers to be experts on the topic, and more important, that they are open to discussing their own risk behaviors at health care appointments, given some reassurances about confidentiality. In this same survey, teens also predicted their parents would underestimate their own child's specific risk: "As predicted by the adolescents, parents in their own focus groups considered drinking a common problematic adolescent behavior, but not typically for their own children."

The Substance Abuse and Mental Health Services Administration (SAMHSA) and American Academy of Pediatrics (AAP) recommend that health care professionals use a screening tool that employs screening, brief intervention, and/or referral to treatment (known as SBIRT) as part of kids' routine health care in order to identify risk behaviors and potential substance use early on and prevent escalation. Screening has the additional benefit of identifying potential drug interactions or contraindications. If, for example, a child presents with symptoms of hyperactivity, it would be important to know if that child is also taking stimulants or some other drug that may interact with drugs for hyperactivity disorders.

According to the attitude survey, when kids understand that the goal of screening is harm reduction and keeping them safe rather than "catching" kids in illegal activities, teens are in favor of screening. Every state has slightly different regulations regarding adolescent and parent rights to confidentiality, so check with the Center for Adolescent Health and the Law (CAHL.org) for details on your state. The limit to physician confidentiality, however, relies on the pediatrician's judgment. If he or she feels that the patient is in imminent harm, or presents an imminent harm to someone else, confidentiality may be broken. In that case, the pediatrician and the patient can discuss how the information will be shared with parents or other health care professionals. According to the survey, "Adolescents often express relief that their parents will be informed of serious problems," and they communicated some relief when they were reassured that they could have some control over how pediatricians disclose relevant information to parents.

Here's how screening, brief intervention, and referral to treatment work:

SCREENING. Many health care providers use a mnemonic, such as HEEADSSS, which stands for home environment, education and employment, eating, peer-related activities, drugs, sexuality, suicide and depression, and safety from injury or violence, to prompt questions about risk factors. Where substance abuse is concerned, screening is meant to elicit information about where a child is on a spectrum of risk ranging from abstinence to abuse.

BRIEF INTERVENTION. The goal of a brief intervention is to prevent, reduce, or stop risky behaviors, and because the health care professional has information at hand about specific risks, the discussion can be tailored to whatever level of use or risk came up in the screening. The brief intervention can range from positive reinforcement for kids who report no substance use, a discussion around the health care risks substances carry for kids who have tried substances but whose use does not rise to dependence, all the way up to motivational interventions for kids who report mild to moderate substance use. For kids whose reported use rises to the level of severe substance use disorder, the physician may opt to take the next step: referral to treatment.

REFERRAL TO TREATMENT. This can take various forms, depending on the severity of use and imminent harm to the patient, but can include referral to counseling, outpatient treatment, or intensive inpatient treatment. Here are some ways to work with your child's physician in order to maximize the benefits of SBIRT:

- ASK YOUR CHILD'S HEALTH CARE PROVIDER if they use a form of screening as a part of regular well-child checks, and if they don't, find one who does.

- LISTEN FOR ACRONYMS. SBIRT is the umbrella acronym used by the American Academy of Pediatrics for the practice of substance use screening, brief intervention, and referral to treatment, but it can encompass many screening tools that work well for monitoring kids' psychosocial health. Here are some others:

  - HEEADSSS

  - Alcohol Use Disorders Identification Test (AUDIT)

- Alcohol Smoking and Substance Involvement Screening Test (ASSIST)

- Car, Relax, Alone, Forget, Friends, Trouble (CRAFFT)

- PROTECT KIDS' PRIVACY AND TIME ALONE WITH HEALTH CARE PROVIDERS. Give your child, and especially your adolescent, the privacy they need to answer screening questions honestly and make sure they have time alone with their primary caregiver to ask questions they may not want to bring up in front of their parent. The American Academy of Pediatrics recommends that children spend some time alone with their health care provider beginning around age eleven to allow for this kind of confidential discussion.

School Nurses

Consider your child's school nurse to be another resource for physical and mental health support as well as another source of screening for your child. School nurses are often the first health care provider to respond to emerging health issues, and they spend about a third of their time attending to student mental health issues. School nurses provide important education and intervention, and after the Centers for Disease Control added opiate overdose prevention to the top five list of public health challenges, the American School Nursing Association came out in favor of training school nurses in the use of naloxone for overdose treatment in schools. School nurses are well versed in the available community resources and can refer students and parents to higher levels of intervention and counseling that may be available.

School Counselors

School counselors are often overlooked (and usually overworked) resources for parents and kids at risk, as they can provide in-school counseling, refer students to community-based mental health services, and collaborate with parents to offer strategies and resources. For families who lack insurance or access to regular health care, school counselors can help families find sliding-scale or low-cost providers and smooth the transition between school-based counseling and more intensive treatment. While counselors don't conduct therapy for kids, they do teach kids coping skills, social skills, and informal accommodations to help kids succeed psychologically, socially, and academically at school. Author and middle school counselor Phyllis Fagell views counselors as an indispensable resource for parents, teachers, and students: "They can give kids strategies that help them feel ready to learn, whether they equip them with mindfulness tools or challenge their faulty, all-or-nothing thinking, or simply provide a safe space for kids to center themselves. Most importantly, they can be the kid's 'person' in the building, that trusted adult they can turn to when they feel distressed or overwhelmed. Every child needs that, but at-risk kids even more so."

Mindfulness

I recently listened to a podcast featuring guest Daniel J. Siegel, clinical professor of psychiatry at the UCLA School of Medicine. He was on the podcast to discuss his new book, *Aware: The Science and Practice of Presence*, and to talk about the benefit of meditation for people in substance abuse recovery, people who, in Siegel's opinion, tend to suffer from a lack of "integration" in their thinking.

This fundamental principle of Siegel's work states that the brain functions best when the higher-order thinking of the frontal lobe and the lower-order thinking of the primitive lower brain are integrated. It is, in his words, "the way we avoid a life of dull, boring rigidity on the one hand, or explosive chaos on the other."

Anyone can lack integration between the rational, higher-order and primitive, emotion-driven areas of the brain, but this lack of integration is more likely for adolescents, whose brains develop on two very different timelines. The emotion areas are in overdrive well before the reason regions mature. Their brains can swing wildly and unpredictability between these states, and as infuriating and perplexing as the swings can be, they are developmentally appropriate. Some teens, however, have more trouble moving toward integration than others, and it's important to address this developmental delay, because lack of control over impulsivity and emotion is associated with early-age onset substance use disorder.

One way to treat a lack of integration, argues Siegel in his books *Mindsight: The New Science of Personal Transformation* and *Aware: The Science and Practice of Presence*, is through a regular mindfulness practice. Multiple studies on mindfulness have found measurable change in brain structures and activation in people who meditate, changes that are of particular benefit in the adolescent brain. The studies describe increased activity in the cortex (especially the prefrontal cortex), insula, and hippocampus, and increased connectivity between the amygdala (remember, this is the highly active emotional response center) and the prefrontal cortex (the higher-order thinking that's still under construction in teens). Integration, or connectedness between these areas, is the holy grail for adolescent cognitive development. It's what allows teens to not only think before they leap, but to weigh the pros and cons of that leap, contemplate the consequences, and plan ahead for the most successful leap possible.

Mindfulness practices promote integration and allow adolescents to focus and gain objectivity and control of their impulses and emotions. Better yet, mindfulness practices can lower the body's autonomic response to stress. Kids who are freaking out over a test or a perceived social media slight may feel physical symptoms such as racing heartbeat, stomachache, cold fingers and toes, or lightheadedness. Having perceptual distance from feelings not only allows them to feel less out of control, but it also lessens the physical responses that make them feel so panicky. Some studies report that mindfulness practices can result in increased "interoceptive awareness," or sensitivity to sensations that originate in the body. If a teen can go from feeling overwhelmed by a stressor, to observing its existence from a bit of a distance, she can begin to understand those stressors as passing experiences rather than overwhelming, out-of-control bottomless pits of despair. Most significantly, this shift in perspective can grant kids the ability to let go—of obsessions, of bad relationships, of material pursuits, or simply of the niggling problems of the day. Depending on the practice, it can also lead to greater spirituality, which can give kids hope and faith in a power greater than themselves.

I know. It can be really difficult to make a case for mindfulness practices, which usually mean meditation or yoga, because the very word with its mom-yoga connotations begs for teen mockery. Here's the thing, though: it works.

I've seen meditation work wonders on my students in rehab, and even the doubters and mockers admit that meditation, as well as yoga, allowed them to quiet their brains and gain some perspective on their chaotic thinking. One of my students pointed out a quote from the book *Spiritual Graffiti* by musician, yogi, and author (and, by his own description, former juvenile delinquent) M.C. Yogi, claiming it captured exactly how the rehab's yoga class makes him feel: "For the longest time in my life, I'd felt like my body was going in one direction while my mind was moving in another. I'd never experienced the serenity of having every side of myself working together, focused and single-pointed [. . .] the energy of being fully present, fully awake, and alive."

After much negotiation, and some cajoling, I convinced Finn to give me just ten or fifteen minutes a day, for thirty days. It helped that he admires M.C. Yogi's music and approach to mindfulness, and I was not above using this information to my advantage. Every day for what I must admit ended up being less than a month, we used Siegel's free "Wheel of Awareness" guided audio meditations. I loved it, of course. I get ten or fifteen minutes of mind-meld with my kid, and maybe—maybe he gains some neurological benefit, some perspective on his emotions, and a little boost to the "protection" side of his substance abuse risk scale. There were no fireworks or magical breakthroughs, but when I asked how he liked it, he said, "It was okay," which, as any parent knows, is akin to a five-star review.

Here are some ways to promote mindfulness practices with your adolescent:

- YOU FIRST. Find a comfortable, quiet, distraction-free place to sit. Pay attention to your breath, in through the nose, out through the mouth. Belly and chest expanding, belly and chest contracting. Whenever you feel centered, take a moment to pay attention to what you feel. Don't worry about fixing it, just notice it. That's it. Welcome to mindfulness practice. Noticing a feeling is the first step toward integrating your perception and awareness (higher-order thinking) with the emotion (lower-order brain function).

- ONCE YOU'VE GOT THE IDEA, INCLUDE YOUR LITTLE KIDS. The younger they start, the more likely they will be to incorporate these practices into their daily life as they grow. The practice is going to have to be fun, brief, and relevant to images and activities that make them happy. First step: find a quiet spot to sit down together, sitting in a comfortable position. Tell your child to imagine herself in a quiet, beautiful place, like a garden or the beach. We tried this practice together even though Finn is older, and he told me he imagines himself near the ocean, sitting on the huge granite boulders that overlook the water, a place that's always made him feel calm. Tell your child to find something in the landscape—one flower, one shell on the beach, or in Finn's case, one wave. Take a deep breath in through the nose and look at that thing closely. Smell it if it's a flower, hold it in your hand if it's a shell, watch the wave break on the rocks, and breathe out, letting bad feelings or stress or anxiety (whatever age-appropriate term you feel will work for your child) go with the breath. Once she's done this a few times, have her open her eyes and tell you how she feels. Connecting how her body feels to words is a powerful act of integration that helps connect feelings (lower brain) to observation and words (higher brain).

- ADOLESCENTS GET MORE CHOICES. They can use either the basic or little kid method, or try one of the ten bajillion other techniques out there. I happen to like Siegel's "Wheel of Awareness," which Finn appreciated because Siegel's voice, rather than his mother's, directed our sessions.

- THEY WON'T TRY IT? Ask for a limited trial period, or if all else fails, offer a trade. One week of daily, ten-minute sessions together in exchange for a dinner or a movie together. I don't often advocate for extrinsic motivators because the evidence on that practice is crystal clear, and they don't work over the long term. They do work, however, as a one-off, to boost motivation or get someone to try something for a short time. Offer an activity, especially one you can do together, rather than a thing like candy or yet another piece of electronics. If your child or teen does not keep it up after that first week, fine, but you have given them the basic tools and an idea of what mindfulness is about, and they may just try it on their own when you are not looking.

Reframe Stress

When I interviewed Dan Siegel about the benefits of mindfulness for adolescents, I asked him specifically about the stress of our recent transition, because I'd been freaking out about the degree to which our move might affect Finn and how I could mitigate some of that risk. Siegel told me I was thinking about our move all wrong. Since adolescents crave novelty and new sensations, why not frame the move as a good thing, as a way to turn the risk factor of risk seeking into a protective, positive outlet for his natural adolescent impulse? A move is a veritable cornucopia of new, exciting experiences with the potential to fulfill Finn's developmentally appropriate desire for novelty, he explained. "Anything that might get the baseline levels [of dopamine] elevated, may be especially sought over," Siegel said, and that's when my head exploded. My fears and worries about the move blinded me to its inherent opportunity as a wealth of protective experiences. According to psychologist Lisa Damour, author of the book *Under Pressure: Confronting the Epi-*

demic of Stress and Anxiety in Girls, "Somewhere along the line we got the idea that emotional discomfort is always a bad thing. This turns out to be a very unhelpful idea."

We can't remove all stress from kids' lives, nor should we, because some kinds of stress can be beneficial for physical and psychological health. So-called positive stress is a normal and important part of growing up and figuring out who we are, usually an experience that is challenging but limited in duration. Think about a time your child was allowed to struggle with a task and emerged out the other side with the words "I did it!" These moments of good stress can feed a child's sense of purpose, self-esteem, self-efficacy, and agency, and are a powerful tool in combating learned helplessness, depression, and addiction. "Working at the edge of our capacities often extends our capacities, and moderate levels of stress can have an inoculating function which leads to higher-than-average resilience when facing new difficulties," says Damour. The second kind of stress, "tolerable stress," is more serious, but it, too, is temporary. Tolerable stress is caused by experiences such as divorce, the loss of a family member, or dealing with a serious injury. These stresses can be alleviated by positive relationships with family or adults that can help kids cope.

Here are some ways to help kids reframe their stress:

- HELP THEM CONTROL WHAT THEY CAN, AND LET GO OF THE THINGS THEY CAN'T. Yes, I am invoking the serenity prayer, but it really does work. "For difficulties that cannot be changed, research shows that practicing acceptance is the critical first step," Damour writes. "Taking a strategic approach—fixing what we can and finding a way to live with what remains—makes it possible to feel less helpless and more relaxed, even in the face of substantial adversity."

- PRACTICE EMOTIONAL WEIGHT LIFTING. When Damour counsels
  students, she offers them a metaphor: learning is like weight
  lifting in that the benefits (whether bigger muscles or more intelligence) happen when the work is difficult. It helps them accept challenge as a part of learning and directs the focus away
  from the stress and toward the potential rewards.

Get Plenty of Sleep

When I tell adolescents how much sleep they are supposed to get,
they laugh. Whole auditoriums full of middle and high school kids
just throw their heads back and laugh. Their message is clear: teens
get nowhere near enough sleep.

Adolescents should be getting between eight to ten hours of
sleep a night, but four out of five teens are getting less than that.
National Sleep Foundation surveys reveal that the average weeknight sleep hovers between seven to seven and a half hours over
the course of adolescence, mainly because teens face two hurdles
in getting a good night's sleep. First, adolescent brains function on
a circadian delay. When teens say they are not sleepy at midnight,
they are not lying. Teen brains simply don't release the hormones
that promote sleep until later in the evening. Adolescents' natural
time to fall asleep is 11:00 p.m. or later, and given early school
start times, there's no way they can get enough sleep. Fortunately,
school districts have begun to shift start times in deference to the
evidence. The American Academy of Pediatrics recommends that
schools start no earlier than 8:30 a.m. in order to prevent chronic
sleep loss in adolescents.

Alcohol abuse is more prevalent among people who get too
little sleep, and lack of sleep causes many of the mood disorders
people tend to self-medicate with alcohol. The link between sleep

and substance use is what's called "bidirectional," in that sleep disorders can cause substance abuse and substance abuse can cause sleep disorders. Alcohol is a sedative, and may appear to work as a sleep aid, but it disrupts sleep cycles because as it's processed, it stimulates arousal centers in the brain. People who drink report falling asleep well but waking up in the middle of the night and being unable to go back to sleep. That pattern was a familiar one for me, and included acute anxiety attacks upon waking.

After a night of poor sleep, teens will be less likely than a young child or an adult to perceive how tired they are. This reduced perception can lead to all kinds of unfortunate situations, from falling asleep in math class to falling asleep behind the wheel of a car. Chronic sleep deprivation is a serious problem for adolescents, and proper sleep habits should be parents' top priority.

First, sleep keeps us physically healthy. People who get inadequate sleep have weaker immune systems, are more likely to be obese, have type 2 diabetes, develop hypertension and heart disease, and have shorter life spans. As a cherry on top, mention that the National Sleep Foundation cites lack of sleep as a cause of acne and other skin problems. Second, sleep improves mental health. People who get inadequate sleep report poorer mood, motivation, judgment, perception of their environment, and functionality for routine tasks. One night of poor sleep can cause crankiness, but chronic lack of sleep has been shown to induce serious mood disorders. Finally, sleep plays a vital role in learning. It primes the brain to acquire and store more information before we learn, and boosts memory consolidation afterward. I like to remind kids that all things being equal, an extra hour of sleep is more effective than an extra hour of studying if they want to do well on that test tomorrow.

Here are some tips for helping your child get enough sleep:

- MAKE SLEEP A FAMILY PRIORITY. If parents don't prioritize sleep, neither will kids. My kids have grown up understanding that in our family, we are flexible about a lot of things, but sleep is not one of them. We schedule around sleep, not the other way around. We also have "quiet time" in the afternoon on weekends that is for quiet pursuits that often include napping. One of our kids has not been a napper since he was a toddler, but he uses that time for screens or reading or music, and respects our family priority around rest.

- SET THE STAGE. After dinner, we encourage (but as our kids get older, don't always prevail) that screens go dark, music comes down a notch, and the house settles at least an hour before bedtime. We prep ahead of time for the things we will need in the morning (pack backpacks, make lunches, check the calendars for things we may have forgotten, such as permission slips) and head to bedrooms staged for sleep. One of our kids likes blackout curtains, the other needs white noise, and both fill water bottles before they head up.

- BE CONSISTENT. Catching up on weekends does not make up for chronic lack of sleep during the week, and worse, too much sleep on weekends can further exacerbate out-of-whack sleeping schedules. Toward the end of summer vacation, prepare to shift bedtimes a few weeks before school starts and keep to that schedule even during vacations.

It's impossible to know which, if any, of these protections will keep my children safer from substance abuse, but I do know that all of the protections I've mentioned have other benefits for our family. Increasing communication and trust is always a goal of mine, and the frank discussion about my own substance abuse, their inherited risk, and how the skeletons in our extended family's closets have harmed relationships is an important part of that. Asking my younger son's physician about her use of screening helped strengthen my trust in her and her understanding of my priorities when it comes to substance abuse, and that's been good for our relationship. I make sure my son has time alone with her at his appointments, because he may want to talk to an objective expert about his risk behaviors and health concerns before he talks about them with me. I made a point to meet the nurse in his new school, just to make contact, so she knows my name and face and may be more likely to get in touch if she notices any issues emerging with Finn. The mindfulness experiment was a bust, but Finn and I tried something new together, and he learned some new skills, so something good came out of that practice even if he did not become a habitual meditator. We continue to work on self-efficacy, as that's a daily exercise. My first book, *The Gift of Failure*, is almost entirely focused on improving kids' feelings of self-efficacy, so my sons are used to my pushing them to try new things and discover how effective they can be in the world. Sleep remains a priority in our house, and as the dogs are more likely to let us sleep at night when they are tired, they help us get more exercise.

Will tonight's dinner together or this afternoon's walk with our nearly deaf, toothless emotional support pug keep my kid off drugs? Who knows, but I can report that on yesterday's hike, Finn mentioned that he's happy here in our new home, and he's actually looking forward to his second year at his new high school. As we sat together, picking wild huckleberries and looking out over the Vermont landscape, I felt the weight of his risk factors ease from my shoulders, at least for a while.

CHAPTER 6

HOUSE RULES
Parenting for Prevention

Nearly every Sunday night since I got sober, I've attended a twelve-step recovery speaker meeting in the basement of a Vermont church. The hour is split between two speakers who share their stories about how their drinking and drugging started, what life was like for them when they were using, how they got to recovery, and what life is like for them now. Some meetings are a master class in storytelling, others less so, but I love every one. I sit in the same place I sat at my first meeting, the one I wept through, with my peppermint tea (the coffee really is terrible), my knitting, and an emergency package of tissues for other freaked-out newcomers.

My favorite part of meetings are the stories. I listen for the common threads, especially at the very beginning when the speaker reveals how they got started with drugs or alcohol. Many of the people in the room are there for the dramatic ending, the climactic spectacle of the crash-and-burn bottom, but I'm there for the quiet parts at the beginning, the origin stories. I want to know what events led up to their first use and what they hoped to get out of that first sip or puff. I want the benefit of all that hindsight now and use it to inform my prevention efforts at home.

Because every substance abuse story begins at home.

Family factors, above all else, are the most important influence on children's well-being and determine their potential for engaging in risky behaviors such as substance use. What happens at home is powerful, even as adolescents begin to turn away from parents and lean toward their peers for counsel. The most significant risk factors for substance use disorder arise out of family factors, yes, but so do the most powerful protections.

Some of the most frequently cited family risk factors include substance abuse in the home, mental illness in the home, domestic violence or sexual abuse, neglect, divorce or separation, adoption, and living in areas where addictive substances are available. Less obvious, however, is the very real risk that lies in not talking substance use with kids. According to Columbia University's Center on Addiction and Substance Abuse (CASA), between 75 and 87 percent of parents talk at least a little about nicotine, alcohol, and marijuana but just 50 to 60 percent talk about other drugs such as heroin, amphetamines, and prescription medications.

Even in the face of these risk factors, family can offer some of the most powerful protections. They can ensure kids have accurate information about their familial risk and the specific dangers of various addictive substances, correct their mistaken and ill-informed beliefs about drug and alcohol use, and help them cultivate healthy, positive relationships.

One quick reassurance before we dive in: many people, including children, can use substances without escalating to the level of substance abuse or dependence. In fact, most do. Addictive substances may be more harmful to children than they are to adults, both because of the damage they do to the brain and because the earlier the first use, the greater the risk of progressing to substance abuse, but lots of kids try drugs and alcohol without falling victim to substance use disorder. That said, it's the kids who use early, progress from occasional use to regular use, and then escalate who are at the highest risk of developing a substance use disorder.

Much of the information kids get about drugs and alcohol comes from unreliable sources, such as social media, other teens, and the internet, but the good news is most teens report getting a significant proportion of their information from parents and school programs. As kids get older, more of their information comes from less credible sources, so it's important to keep accurate information flowing, especially as kids enter high school. One interesting data point underlines the impact of communicating credible information to kids: teens who rely on peers, social media, or the internet for their information on addictive substances were twice as likely to report an intention to use drugs in the future than kids who rely on more credible sources such as parents and school education programs. As we'll see in chapter 9, evidence-based education programs have come a long way from the programs that dared us to stay off drugs and "just say no."

Parenting: A Balancing Act

There is a balance to be struck in parenting for substance abuse prevention: parents want as much information and reassurance as we can get our hands on, but our children want (and need) autonomy, privacy, and to feel trusted. We may tell them we trust them, but when we monitor and surveil kids excessively, we communicate the opposite message. This is where parenting gets complicated, where we have to rely on our honest assessment of our kids' friends, common sense, and tolerance for risk.

On the one hand, there is data that suggests kids who hang out unsupervised with peers who use drugs are more likely to use drugs themselves. The Center on Addiction's 2019 report, *Teen Insights into Drugs, Alcohol, and Nicotine: A National Survey of Adolescent Attitudes Toward Addictive Substances,* concludes that teens with more unsupervised time are more likely to use addictive substances. Sure, we could keep our teens from using substances if we make sure they never have time alone with their friends, but that's like saying parents can prevent all sunburns by never letting their children go outside. It might work, but kids need time alone with their friends just as much as they need to go outside, especially as they enter adolescence, when their need for privacy and autonomy increases in lockstep with their risk of using addictive substances.

This report also urges parents to monitor kids' social media accounts on the logic that "teens who reported spending more time on social networking sites, and especially those who reported seeing pictures on those sites of 'kids drunk, passed out, or using drugs,' were more likely to report engaging in tobacco, alcohol, and marijuana use." Here, again, I urge balance. While parents who monitor every aspect of their kids' digital lives may be more likely to intercept problematic communications, they will also be more likely to intercept innocuous communications that should have been kept private, communications that are vital to kids' friendships or their personal expression.

Parents can also do all sorts of collateral, unintended damage to kids' relationships through excessive monitoring. I spoke with a thirteen-year-old girl I will call Tracy who felt isolated and sad because she'd been cut out of her friends' text messaging group. When I spoke to her friends, they revealed that they knew Tracy's parents monitored her emails, texts, and social media posts (even after her parents ensured the security on Tracy's accounts was set to private) and did not want Tracy's parents knowing their business. They felt their privacy was being invaded, and consequently, cut her out of their virtual peer group.

I am not saying you can't monitor children's digital lives. I'm simply asking you to weigh the pros and cons very carefully. I have never, not once, monitored either of my children's emails, texts, or social media accounts, and I understand that makes me an outlier. According to the Pew Research Center, a surprisingly large percent of parents admit they have checked which websites their kids visit and their social media profiles (including checking to make sure the settings are set to private), looked through their teens' phone calls and messages, and used monitoring tools to check on their teens' location using a cell phone. Adolescent psychologist and author Lisa Damour challenged parents in an article in *Time* magazine to ask themselves: If we are raising one of the best-behaved generations on record, kids who use less tobacco, alcohol, and drugs, have less sex, and are more likely to hang out at home, "why do we worry so much about them?" She concludes, based on her research and years of experience in private practice and working with schools, "Regardless of how closely we decide to monitor our teens' lives digitally, no amount of surveillance can take the place of having a sturdy, working relationship with them."

Put the bulk of your effort toward the relationship, not the surveillance. When a parent and child have a warm, mutually respectful relationship, the child is more likely to perceive monitoring as coming from a place of warmth and respect rather than control and manipulation.

Style Your Parenting for Prevention

In 1989, clinical and developmental psychologist Diana Baumrind announced the results of the Family Socialization and Developmental Competence Project, a twelve-year study aimed at identifying how parenting styles and other factors in family life affect the competence and development of children. Baumrind posited that parenting style has a significant impact on adolescent substance use, particularly two factors: "strong mutual attachments that persist through adolescence, and coherent consistent management policies including supervision and discipline." In the years since the publication of this landmark study, there has been a great deal of debate about parenting styles and whether permissive, authoritarian, authoritative, or some other classification outside of Baumrind's classifications can prevent or precipitate drug use and abuse.

Today, developmental psychologists define parenting styles according to four categories:

AUTHORITARIAN: highly demanding and directive, but not very responsive to children's needs

AUTHORITATIVE: demanding but responsive to children's needs

NEGLECTFUL: neither demanding nor responsive

PERMISSIVE/INDULGENT: not demanding of kids, but highly responsive to their needs

These parenting styles are distillations, of course, and parents can be shadings of each depending on the day or situation. However, some generalizations can be made that are helpful in raising addiction-resistant kids. In a 1991 journal article, "The Influence of Parenting Style on Adolescent Competence and Substance Use," Baumrind concludes, "Secure in their attachment to their parents and with adequate protection from the instabilities present in the larger society, adolescents from authoritative homes showed that they simultaneously could validate the interest of personal emancipation and individuation, and the claims of their shared social norms."

In other words, if you want to raise kids who feel supported and connected to their parents, kids who are relatively protected from substance use and feel a sense of self-efficacy and competence, adopt an authoritative parenting style. Generally, authoritative parents do the following:

- Have high expectations for conduct

- Enforce high expectations fairly

- Are warm and supportive

- Support kids through failures and mistakes

- Are more democratic and offer verbal reasoning for expectations as opposed to "because I said so" dictum

- Prioritize children's learning over blind obedience

- Give children autonomy and space to learn and grow

- Recognize the child's own interests and goals as valid and worthy of support and respect

Here is some advice on the practice of authoritative parenting against the backdrop of the three domains of parenting that have been shown to have a significant impact in preventing substance use and other risky behaviors in children and adolescents: knowledge, expectations, and practice.

Knowledge

Authoritative parents teach and guide children according to a "here's why" approach as opposed to a directive, authoritarian "because I said so" strategy, and as any teacher will attest, you have to understand a topic yourself before you can help someone else understand it. For example, the information on substance use in the context of brain development in chapter 3 can provide the knowledge you need in order to fully explain why drugs and alcohol are especially dangerous in adolescents' developing brains. Knowledge allows parents to offer persuasive information based on facts. "Knowledge" also applies to our understanding of our children, and I'm not just talking about being aware of their whereabouts and the contents of their email. Any parent can monitor a child's location, peers, and activities, but true knowledge requires understanding a child's goals, hopes, dreams, motivations, and fears. This kind of knowledge provides insight into their decision making, which allows parents to be proactive in their parenting, rather than merely reactive.

Expectations

By expectations, I mean family values, attitudes, beliefs, and norms, the standards you hold each other to. These mutually agreed-upon expectations help everyone understand how to act and why, and keep everyone accountable. Authoritative parents explain the reasoning behind expectations, because rules enforced without reason tend to be ineffective, especially as children grow into adolescents and begin to mature into their frontal lobe, higher-order cognitive abilities.

Practice

Practice is where knowledge and expectations join forces and where action meets consequence. Practice dictates how individual members apply family expectations in daily life, and how parents react when expectations fall short. The best parenting practices help children feel an even greater sense of self-efficacy by connecting actions to natural consequences, whether good or bad. Make a good decision, and they get to enjoy the good things that happen as a result. Make a bad decision, and the consequences are theirs to deal with as well. Sure, this is the hard part, but enforcing consequences does not make you a mean parent, it makes you the type of parent who does not expect the world to make exceptions for your child.

The first order of business, then, is for parents to learn as much as possible about how drugs and alcohol affect developing brains and bodies so we can then use that knowledge to educate our kids (hopefully, in lockstep with effective school-based programs), work as a family to establish family expectations for substance use, and enforce those expectations with logical, clear consequences.

An important place to start is to figure out whether you want to be a zero-tolerance family or a family that expects—even allows— kids to experiment at home. Many parents (depending on the survey, between 20 and 50 percent) believe if they let children have sips of alcoholic beverages at home when they are little, it will prevent them from engaging in irresponsible drinking later. In fact, the study "Really Underage Drinkers: The Epidemiology of Children's Alcohol Use in the United States" found sipping at home offers no protection against drinking in adolescence, and it seems to have the opposite effect: kids who sipped in fifth grade were twice as likely as their non-sipping peers to progress to regular drinking habits in grade seven. Despite the folly of this practice, it's not uncommon. The study reveals that among sixth graders, 62 percent of boys and 58 percent of girls had tasted alcohol and 29 percent of sixth graders had consumed "more than a sip," which could mean anything from two sips to an entire glass. Only one study has ever asked younger children about more frequent use, and that study of children in Arizona found nearly 5 percent of kids between third and sixth grade reported having had a drink in the past week. Kids who start early at home are more likely to become regular drinkers by middle school. And what of the pervasive, romantic image of free-sipping European children who grow up to have healthy relationships with alcohol? It's time for that myth to die, because Europeans do not have a healthy relationship to alcohol. According to the World Health Organization (WHO), the European region has the highest alcohol consumption levels, the highest proportion of alcohol-induced negative health consequences, and the highest rate of deaths attributable to alcohol in the world. One-fifth of Europeans aged fifteen or older report binge drinking at least once a week. Half of European males are binge drinkers, and 60 percent of adolescents fifteen to nineteen years old drink. In fact, the French government launched a national campaign in 2019 aimed at reducing alcohol consumption to "a maximum of two glasses per day, and not every day," after WHO surveys revealed France has one of the highest alcohol consumption rates in the world, and drinking is the second-largest cause of death after tobacco. The "Europeans do it" fantasy of raising kids with sips or their own glass of watered-down wine so they grow up to be moderate, social drinkers is just that: a fantasy.

It's not the mouthful of wine or beer that causes children to start drinking early, of course, it's the permissiveness behind it that sets the stage for continued substance use. A child who is allowed to drink understands that drinking is no big deal in their home. And a permissive stance on underage drinking becomes part of family expectation and practice. Candy Finnigan, a substance interventionist known for her work on the show *Intervention*, writes in her book *When Enough Is Enough*,

> A lot of kids now are supplying themselves initially from the liquor at home. Most [peer] groups that party together have worked out which homes are "safe" for drinking after school, whose mom works late, which home is easiest to sneak out of during sleepovers, and unfortunately whose house has a welcome mat outside—where parents drink and drug themselves, allow parties and worse under their own roof, or reason that "I'd rather have them doing it here." I must point out the insanity of such thinking. Any parent with knowledge of underage drinking and drug use is guilty of a felony—contributing to the delinquency of a minor—as well as aiding and abetting a more serious crime.

If you, as the parent, are not comfortable with a total ban on drinking, if you are worried your child will think you uncool, you have an easy out: underage drinking is illegal.

The American Academy of Pediatrics recommends parents and physicians send a "clear message against the use of alcohol by adolescents and young adults under the age of 21 years," and one parent I interviewed outlined his extraordinarily clear messaging in an email.

1. I do not allow any illegal activities, and last time I checked drinking under the age of 21 was illegal.

2. Every adverse outcome that you can measure that happens during adolescence you always find alcohol. Whether it is an STD, an unwanted sexual contact, an accident, a fight, or an overdose; you find alcohol.

3. The other thing I said was should you get invited to a party where adults are serving alcohol, I assure you I will show up with the sheriffs and have them hauled off.

While this parent's expectations are hard-core, they are also logical, and rely on natural consequences, or consequences that would be most likely to happen in the real world. Natural consequences are the most effective way to help kids understand the relationship between cause and effect. This education should start when they are young, so for example, if a young child forgets her lunch when she heads off to school, she will be hungry at lunchtime or have to share with a friend. If a child forgets his math homework, he will get a zero for that assignment. If your teen drinks alcohol when he is under twenty-one, the police may cite him or take him to jail.

Even if you are not willing to go so far as to have your children or other adults arrested for underage use or contributing to the delinquency of a minor, it's important to implement natural consequences from an early age because they are the best possible way to help kids understand the connection between their actions and the effect of their actions. If your child stays up late, he may be tired in the morning, but it's important he get up, and you help him connect the dots between his behavior (not going to bed) and the natural consequence (being tired in the morning). When parents impose punishments for bad decisions or other unwanted behaviors that are unrelated (for example, taking away television privileges when a child won't take a bath) they make no sense to the child and are unlikely to help her learn. Remember, their brains are still under construction, so they are less able to connect cause and effect when they have no relationship to each other. Further, when parents rescue children from the natural consequences of their behavior, for example by allowing kids to be late to school so they can sleep in, kids are less likely to learn from their actions. Understanding connections between behavior and natural consequences will become increasingly important as kids get older and the consequences for bad decisions get more serious, and it's important that parents stay strong, resist the temptation to rescue, and let kids learn from their mistakes.

An easy way to codify family expectations around substance use is to create a family contract. It should be the product of discussion between all family members, based on your shared values. A family contract can also serve as one more way to communicate openly and clearly about your family's expectations and the consequences that will follow if the contract is violated. Here are some sections to consider:

- CLEAR EXPECTATIONS for everyone's relationship to substances. This could include expectations for parents, expectations for children, or ideally both. For example, in our family, no drug or alcohol use is tolerated among underage children; I do not use addictive substances; and when Tim purchases alcohol, he drinks what he wants from the bottle or can and then disposes of the rest so we do not ever keep open containers of alcohol in our home.

- CLEAR CONSEQUENCES for violating family expectations.

- PLANS for scripts kids can use to decline drugs or alcohol when offered away from home.

- HONESTY and a promise to discuss concerns about risks as they emerge.

- AN EXIT STRATEGY in the form of a promise to pick your child up, no questions asked, if they do not feel safe.

Once you have established clear expectations and consequences, there is still a lot to talk about as the family grows, matures, and works through changes to the family structure. Here are the kinds of special considerations that impact family expectations around substance use.

When Addiction Runs in the Family

Experts agree that genetics make up about half of the substance use disorder causation equation. I've heard many experts use the following analogy: if genetics are the loaded gun, then a child's environment, especially adverse childhood experiences, are the trigger. This is incredibly powerful knowledge for parents to have but even more powerful in the hands of kids who will have to make decisions about whether or not to use drugs and alcohol. Knowing whether or not substance use disorder lurks in your or your child's genome, however, requires honesty and open discussion. The elephant in the living room must be identified as an elephant. If you try to pretend it's not there, or that it's something other than an elephant, the inevitable result is confusion and distrust.

Here's what we know about the genetics of addiction. There is no single "gene of addiction," and even where experts have identified specific genetic markers, the picture is much more complicated. The genes that control whether or not a child may be susceptible to substance abuse also control many aspects of personality that can predispose them to just that. For example, the D_2 receptor gene, or DRD_2, is implicated as a source for susceptibility to substance use disorder (SUD) but it may also be a source for various personality traits that render kids more likely to seek out and use addictive substances, such as impulsivity, sensitivity to rewards, sensation-seeking, and novelty-seeking. What's more, it's not just the existence of the DRD_2 receptor that matters, because how a given gene is expressed in a particular area of the brain can also influence behavior. To top it all off, both genetic and environmental factors can affect kids' behaviors differently during various stages of development. Any statements researchers make about which genes affect future substance use or abuse must include this caveat: genes impact susceptibility to addiction and formation of personality, and the two can't be teased apart so easily.

It would be so much easier if researchers could point to particular genes as the cause of addiction, and as gene-knockout technology becomes more refined and accessible, we could knock those genes out, thus eliminating substance abuse forever. We are not there yet. We can, however, use our knowledge of genetic predisposition to shape expectations and practice. If I know my kids are more likely to suffer from substance abuse, their heightened risk can become fodder for communication around family norms. I will provide specific tips and scripts in chapter 7, but from very early on, you can use the risk of genetic predisposition as a potent and effective protective factor.

Even when children don't have a genetic predisposition, or when it's impossible to know what their genetic history looks like, as in the case of adoption, addiction and family history is about so much more than genes, which is where epigenetics and environmental forces come into play. Children's perception of alcohol norms, such as drinking habits and frequency of use, come from observing their parents, and this education begins at a very early age. In one study, children as young as three were able to identify alcoholic beverages in photographs, and children of alcoholics were able to identify more alcoholic beverages than children without alcoholic parents. The authors conclude that children begin to internalize cultural rules (including family rules) around drinking alcohol before entering preschool. Modeling a healthy relationship to drugs and alcohol, whether that looks like abstinence or moderation, may just be one of the most important lessons we teach our children.

Families with at least one alcohol-dependent parent also tend to have more negative family dynamics and are less likely to solve problems effectively. This effect is heightened in families where a parent engages in binge drinking. When home life is traumatic, chaotic, or stressful, kids suffer on many levels. Kids learn how to regulate their own behavior by watching the people around them, so when kids are raised by adults who act erratically, fail to support or protect them, or who can't manage the stresses in their own lives, kids are more likely re-create these behaviors as they develop. Children of alcoholics also have steeper and more rapid escalation rates of use than children of non-alcoholics, experience consequences of their drinking earlier, and are more likely to develop alcohol dependence than the children of non-alcoholics.

Talk openly about your family history of substance use disorder and teach your children about the impact of their genetics. Help them understand that because grandma or dad or their sister has a substance use disorder, they are much more likely to become dependent on drugs and alcohol. This is easier said than done, as substance use disorder is weighed with shame so much so that some people go to their grave never having said the words out loud. The first time I admitted I had a problem, I threw up. The second time, I cried, but I managed not to barf. Every single time I've talked about it since, the words have come easier and easier. These days, I have a T-shirt that announces my sobriety in giant letters across my chest, and I wear it when I need an extra boost, a reminder that I've got this whole sobriety thing under control for the day. Whenever I wear it, or mention my sobriety publicly, good things seem to happen. I've been in a position to help other people, including my children, because I talk openly about my own substance abuse. Make it clear that your family expectation is that alcohol use before age twenty-one will not be tolerated, not just based on state and local laws, but on your family's heightened risk. The stakes are higher for kids with a genetic predisposition for substance use disorder, so help them use that knowledge to guide their decision making.

Siblings Who Use

Siblings don't just share genetic material; they also share attitudes and perceptions around alcohol use and can be powerful influences for their brothers and sisters. Research shows older siblings contribute to family expectations around drug and alcohol use, so even when parents are clear on family expectations, siblings have the power to strengthen or undermine these norms. Younger siblings often look up to their big sisters and brothers, and may be more likely to emulate them, for both good and bad. When kids know—or even if they mistakenly believe—an older sibling is using drugs and alcohol, they will be more likely to use drugs and alcohol themselves. On a practical level, older siblings often serve as younger siblings' first supplier. Surveys show one out of ten teens who use addictive substances report that it was their older brother or sister who encouraged them to try it or who supplied the substance itself.

Older siblings wield a lot of influence over younger siblings, and not just around substance use. If older siblings have warm, supportive friend groups, younger siblings are more likely to emulate that dynamic. Use this knowledge. Capitalize on a child's sense of responsibility and protection for their younger siblings, and recruit them into the family's substance use prevention effort. Help them understand they are teachers for their siblings, often as important as a parent when it comes to modeling behavior and decisions. Mentoring younger siblings can have the added effect of causing older kids to rethink their own substance use. It's amazing what some responsibility and leadership can do for a kid's sense of duty to others, not to mention his morale and self-esteem.

Personality Traits

People in recovery often talk about finding their "drug of choice," the substance or category of substances that really does it for them. Some describe it as a "click," a sudden feeling of "ahhhhhhh," or the filling of a hole. In the book *High*, by David and Nic Sheff, Nic explains why meth was his drug of choice: "As soon as the drug hit me, I felt a rush of elation—not just from the drug, but from feeling like this was what I'd been looking for my whole life. It was better than those first hits of pot, better than *everything*. I felt super confident, super strong. I felt like a real-life superhero. Just like that, I was addicted."

Nic's personality fit with meth. Others may find what they've been looking for in alcohol, or pot, or cocaine. Whether it stems from personality type or the shape of a kid's particular need—for love, for more confidence, for a sense of self-efficacy or worth, or for relief from emotional pain—knowing your kid can help you know how to help them find what they are looking for without having to turn to chemicals.

Certain personality traits heighten the likelihood kids will use addictive substances, so knowing about these traits can help parents and teachers identify these kids and offer more support, intervention, and monitoring before use turns to abuse. Some intervention and education programs, such as the Preventure program, use personality testing to identify higher-risk children and shape intervention and treatment. The personality types Preventure focuses on are: sensation-seeking, impulsiveness, anxiety, sensitivity, and negative thinking. Other studies have noted a discomfort with boredom, which may be another trait worth noting, and one study found that people with a tendency to get bored easily are more likely to use opiates more often. One large study of college students concluded that a lack of self-control is another trait that can be predictive of future SUD.

Here's where we run into yet another chicken-and-egg problem. Stressed-out, anxious, and depressed people may be more apt to smoke tobacco, but tobacco use can contribute to a person's stress, anxiety, and depression. People living in poverty are more likely to smoke, because poverty causes higher levels of stress, anxiety, and depression. The entire field of study is a statistical minefield. Finally, it's important to remember that kids are constantly shifting creatures whose personality traits can change with the direction of the wind. Given all of these caveats, knowing your child well can help parents tailor their alert levels when it comes to substance use. Because most kids can try substances and not progress to abuse or dependence, it's helpful to know which kids might need more guidance and support.

The good news is that there are ways to take advantage of teens' heightened sensitivity to rewards, need for novel experiences, risk, and sensation. Encourage kids to try new activities. Talk about ways to channel their developmentally appropriate need for risk- and sensation-seeking. Encourage kids to audition for a play or a solo in the school concert, take up a sport that itches the need for adrenaline and rewards, such as rock climbing or bike racing. Numerous studies show, "being both physically active and a team sports participant was associated with a lower prevalence of several risk behaviors," including substance abuse and risky sex. Exercise also raises endorphins, dopamine, and serotonin, and animal studies have shown that vigorous physical activity can reduce drug-seeking behaviors. Trade in birthday and holiday presents for experiences, adventures that allow for everyone in the family to stretch and risk and grow together, opportunities to spend time away from home.

Divorce, Separation, and the Single Parent

Divorce and separation are so common in today's society that we often forget these events can have a significant impact on children's emotional well-being. Divorce is, according to one researcher, "the most prevalent of adversities experienced by adults and children in the U.S." Between 40 and 50 percent of marriages established in the 1990s will end in divorce, and roughly one-third of all children in the United States will experience parental divorce before their sixteenth birthday. No matter how hard divorcing parents work to keep the transition smooth, amicable, and gentle, the experience of divorce is, at baseline, a stressful and confusing experience for kids. For the sake of discussion, I will lump separation and divorce together, because the end result—parents living separately with the subsequent geographical, emotional, and educational upheaval that typically follows—is similar when viewed through the child's experience.

Divorce is included on the CDC's Adverse Childhood Experiences quiz because, no matter how amicable the split, it's a stressful ordeal. When divorce is contentious and chaotic however, it can be deeply traumatic for children. A large meta-analysis initially reported in 1991 and repeated in 2001 showed that "children with divorced parents continue to score significantly lower on measures of academic achievement, conduct, psychological adjustment, self-concept, and social relations." Often, divorce significantly changes the custodial parent's financial situation, and single-parent households subsequently have more limited financial resources than two-parent households. Given the rate of divorce isn't likely to fall, it is essential that parents acknowledge the stress divorce places on the children involved and put as many protective factors in place as possible to counterbalance the traumatic aspects.

It's important to note that the period prior to divorce and separation is a risky time for kids as well. Several studies have shown teens may increase their substance use during this period in order to cope with the stress of heightened conflict. Children who display negative feelings regarding their parents' divorce have higher levels of depression than their peers, and of course, if the divorce or separation happened because one or both parents have their own substance abuse issues, the kids are at greatly increased risk of turning to addictive substances themselves.

The good news is that there's a lot we adults can do for kids who are going through this experience. Support, both from parents and peers, is key. Kids of divorce need to have someone they can talk to about how they are coping, someone who will not be defensive or shame kids for having intense feelings about the situation. If neither parent can provide this, an objective, trained, third party can be a powerful ally for a child of divorce, whether that's a counselor, court-appointed advocate, guardian ad litem, teacher, pastor, rabbi, or some other trusted adult capable of listening and supporting.

When a troubled marriage ends in a respectful manner, research shows, it can provide relief for the entire family and act as a protective factor for kids. Strong connections between parents and kids will be an essential element of any healthy divorce or separation. Healthy, consistent, trusting connections are always important, but during a difficult transition it's even more important that kids feel as if they can share their concerns and fears.

Some divorcing parents struggle with loss of control during a divorce or separation, and subsequently exert more control over their kids as a way to make up for it. Remember that reasonable monitoring of kids' internet use, education, and physical whereabouts can be mutually beneficial, but excess control makes them feel distrusted, helpless, and resentful. Strike a healthy balance between trust and control. When parents manage their own stress around divorce and separation, children benefit, and when children of divorce are given sufficient support, love, and outlets for their stress, their risk of turning to addictive substances is reduced.

Teens are at the highest risk for substance abuse leading up to and during separation and divorce. As time passes, the risk drops.

The most effective and urgent interventions should be implemented just before and at the time that separation and divorce take place. One program, New Beginnings for Divorcing and Separating Families, has been particularly effective in promoting resilience for children of divorce. A group-based intervention, it focuses on helping mothers boost their children's interpersonal skills, implement effective disciplinary strategies, strengthen parent-child relationships, and reduce conflict at home. Studies of New Beginnings show that after completing the program, mothers improve their coping skills, their children report less aggression, and the parents are more open to discussing visitation and custodial arrangements.

Divorce or separation can leave kids feeling helpless, powerless, and out of control. Research on learned helplessness shows that the best way to neutralize and reverse such feelings is to give kids more control and autonomy, even as the world around them changes. Let them choose what they help make for dinner, how their room is decorated, how their schedules are structured. Whenever possible, keep kids connected to their friends, activities, and sports teams. As research shows, even when homes change, stability at school can help kids feel a sense of continuity, so try to maintain as much continuity in school and school-based activities as possible.

In the midst of all the change divorce brings, kids can feel disconnected from their parents, so ramp up your efforts to connect. Protect routines, especially dinnertime and bedtime. Cook dinner together when you can, and work together to pack lunches. Support kids in their efforts to establish new routines, schedules, and habits in their new homes, and ensure they have everything they need if they are spending time in the other parent's home.

Finally, remind them, over and over, and then once again, that family expectations have not changed just because their parents are no longer living together. Consistent rules are as essential as consistent love and support, and can provide an anchor for kids as they weather the storms of family separation.

Substance Use and Gender Differences

Before we dive into the differences between girls and boys when it comes to substance abuse, we have to talk about what sets kids up to drink or drug. Trauma and other negative life experiences, as we have discussed, set kids up to self-medicate, and this is particularly evident when we look at the data on gender and addiction. For example, sexual abuse is one of the most significant risk factors for early substance use, and girls are more likely to have suffered childhood sexual abuse than boys. Their risk increases as they age, and abuse increases risk for developing substance abuse over a girl's lifetime. In a 2014 study published in the *Journal of Adolescent Health*, 16.8 percent of fifteen-year-old girls, 21.7 percent of sixteen-year-old girls, and 26.6 percent of seventeen-year-old girls report they have been sexually victimized by an adult or another adolescent sometime during their lifetime. The rates are much lower for boys (4.3 percent, 3.3 percent, and 5.1 percent, respectively). On a related note, parental alcohol abuse increases a child's risk of experiencing sexual abuse.

Alcohol

In adults, men drink at higher rates than women, and are more likely to binge drink. However, in adolescents, the picture is flipped: while boys start drinking alcohol earlier than girls, girls drink slightly more once they get started and are more likely to binge drink. As girls and women have different body compositions than boys and men (more fat, less muscle, and therefore less water), and less of an enzyme required to break alcohol down in the body, they metabolize that alcohol less efficiently than boys. This means girls and women generally get drunk faster, stay drunk longer, and feel worse on the morning after than their male counterparts who drink the same amount.

If they keep drinking into adulthood, it will make them sicker, faster. Women have higher rates of liver disease, stroke, and heart disease as a result of their drinking and are more likely to suffer drinking-related accidents, and die by suicide. Of course, the ill effects of drinking are not limited to chronic illness. Girls and women who drink heavily are at much higher risk of being the victim of violence including sexual assault and of having unprotected sex (and suffering all the consequences that may entail, including unwanted pregnancy or acquiring a sexually transmitted disease). Women can't even revel in the news that a drink a day may be good for your health. Recent research shows as little as one drink a day can increase a woman's risk of breast cancer, especially when breast cancer runs in the family.

Marijuana

Boys and girls experience slightly different highs, and different cognitive deficits during that high, when it comes to marijuana. Some of this may be due to the fact that girls have more finely tuned executive function skills to lose when they get high, especially during late adolescence. As I've mentioned, that is because girls' gray matter volume increases earlier and peaks about two years before boys' on average. Because pot-induced executive function deficits are greatest in adolescents with greater prefrontal cortex volume, girls have further to fall when high than their male counterparts.

Men are more likely to be addicted to marijuana, but women are more likely than men to suffer from panic and anxiety disorders as a result of their marijuana use. Women progress more rapidly in their substance abuse than men, prompting some substance abuse researchers to conclude that women are more vulnerable to the negative consequences of addiction than men. This could be good, in that women tend to progress to treatment more rapidly than men, but it also implies the factors that often lead people to seek out treatment—a "bottom," as it's known in the twelve-step world—may occur more rapidly among women than men, although relatively few pot abusers seek treatment when compared to abusers of other drugs.

Stimulants

Women are more vulnerable to stimulant abuse for biological, mental, and cultural reasons. In animal studies, females take to cocaine use faster, and take more of it, than males, possibly because females may be more sensitive to the rewards stimulants offer. One small study even showed women reported better-quality highs when they smoked cocaine during particularly progesterone-saturated phases of their menstrual cycle. Women are more sensitive to the cardiovascular effects of cocaine, but female brains seem to suffer less damage from cocaine abuse than men's.

Just as women in the 1950s took doctor-prescribed amphetamines for weight loss and increased energy, today's women report they use methamphetamines to cope with the demands of work, child care, and responsibilities at home, and girls begin taking methamphetamines earlier than boys do. Adolescent girls are far more likely to diet than their male counterparts, and diet pills are a tempting shortcut for fast weight loss. Nearly 13 percent of teen girls admit to using diet pills as compared to 5 percent of boys. Girls who diet are more likely to smoke and drink, and when they do, they smoke and drink significantly more than girls who don't diet, even if those dieting girls are not taking speed to lose weight.

Our family line on alcohol and drugs has been a moving target since the day my children were born; all I have to do is look at my older son Ben's baby book for proof. In the first weeks after his birth, I affixed the label from the first wine he ever tasted, a 1990 Château d'Yquem Sauternes. A close family friend sent this outrageously expensive wine to celebrate Ben's birth, and by our thinking, it made perfect sense to make sure the first wine Ben tasted would also likely be the best he'd ever encounter. I'm sure infant Ben appreciated the "intense spice, honey and dried orange peel aromas" that *Wine Spectator* noted in their review of the vintage.

We've let both boys sip. We've allowed them to watch films that romanticize drugs and alcohol. And I have exposed them to the chaotic and unpredictable behavior of the alcoholics in my family. Worst of all, I spent their formative years drinking all the wine I could get my hands on. They claim not to remember seeing me drunk, but they definitely saw me drink. I have made a lot of mistakes.

All I can do today is move forward and make the next right decisions based on the knowledge I've gained. I've shared much of what I have learned about substance use and the adolescent brain with my boys, and together, we have shaped our current family expectations based on that knowledge. There have been natural consequences when those expectations have not been met, consequences that they may not have enjoyed, but they definitely understand. We talk and we talk and we talk, and every day it gets easier to move past my shame and regret and focus on what I'm doing right today. Ben won't remember his first taste of Château d'Yquem, but I do hope he remembers his mother was honest about the things she got wrong, learned from her mistakes, and expected no less from him.

CHAPTER 7

WE HAVE TO TALK ABOUT IT
Starting the Conversation

When my boys were little, they told me everything. I knew about their likes and dislikes, dreams and nightmares, triumphs and failures. It was all so easy to access, right there on the surface. As they got older and began to develop their annoying but completely developmentally appropriate spheres of privacy, my husband and I had to get more creative. For years, the usual parental tricks worked. Tim conducted sex talks on the Dartmouth Skiway chairlift with my goggled and helmeted tweens, while I preferred the idle time of long car trips. After fifty miles or so, defenses melted, and conversation could expand to include topics that usually hide in wait behind the small talk. These days, my boys are more likely to take ski lifts and road trips with their friends, so it's been much more difficult for us to gain access to their private lives.

As our relationships mature, the one place everyone's defenses predictably relax is at dinner. When the boys were little, I knew I wanted our dinner table to be the gathering place in our home. I sold my grandmother's delicate, heirloom drop-leaf table in favor of something more appropriate for a home filled with children and large animals (my exact words to the builder were, "I'd like a table we could tap-dance on, please"), and that sturdy table has been at the center of our family for nearly twenty years. Most dinners have included a recounting of our "high, low, funny," the best, worst, and most entertaining moments of our day, and a game of *Tier auf Tier*, in which we take turns stacking colorful wooden animals into precarious towers. I always managed to come up with a unique high from the day, but in truth, my high was always the same: dinner, together, hearing about their best and worst moments, while balancing my penguin atop Ben's monkey and Finn's sheep.

Now that Ben goes to college an hour away from home and Finn has a part-time job, our family dinners are less frequent and more disorganized. I'm not going to lie: dinner often takes place around our kitchen counter rather than at the table, and there's often a food documentary or political comedy talk show playing in the background. Tim and I merit smaller and smaller helpings of our boys' time and attention, so we have had to make compromises and learn how to adapt dinner to work for all of us.

No matter what our dinnertime looks like, I continue to protect it whenever possible. According to Joseph Califano, founder of the National Center on Addiction and Substance Abuse (CASA), having dinner together as a family is one of the most important things parents do to protect kids from all kinds of high-risk behaviors, including substance use. Dinnertime is valuable not because of our sturdy table, the food we cook, or even the little wooden animals, but because we predictably and regularly come together to talk, report back on life outside our family, look each other in the eye, and take each other's emotional temperatures. If someone in the family is bummed out or elated we will know, and I'm not just talking about the kids here. If one of the boys could not meet our eyes or seemed disconnected, we'd know, and if I started drinking again, if I could not meet their eyes or I disengaged from them emotionally, they would know that, too.

According to CASA surveys, the more dinners per week kids have with family, the less likely they are to drink. Thirty-three percent of kids who eat zero to two dinners a week with family use alcohol, but if kids eat dinner five to seven times a week with family, the prevalence rate drops by half. No, it's not quite that simple, and yes, there are plenty of confounding statistical factors at play in the research on substance use: kids who are more likely to have dinner with their family are also more likely to have an intact family, synched schedules, and all the other advantages a stable family life implies.

After our move to Vermont, I renewed my efforts to keep dinnertime at the center of our family life as an anchor at the end of our increasingly complicated days. I texted enticing menus to Ben at college in an effort to lure him home. We made menu compromises to please both the newly vegetarian members of the family and the perennially meat-loving son. My husband shifted his schedule so he could get home from work earlier in the evening. It worked for a while, but as life got more hectic, family dinners became an exception rather than the norm. And then, during one of our pizza-on-the-countertop-with-a-side-of-YouTube dinners, I was struck with an idea.

One of our family's favorite talk shows is *Hot Ones*, hosted by Sean Evans. Evans interviews celebrities while they eat a series of ten chicken (or vegan) wings, each coated in a different hot sauce. He starts at the low end of the Scoville scale (a measurement of spiciness) and ramps up. The first wing is mild, at around 1,500–2,000 Scoville units, and the final wing, complete with a "last dab" of extra sauce, clocks in at 3.3 million Scoville units. Evans is a great interviewer; he is well prepared with extensive research, but something about the escalating spice knocks his guests off balance, revealing their deeper vulnerability and truth along with a lot of tears and snot.

I sent a request to our family text group to secure a date for a VERY SPECIAL LAHEY FAMILY DINNER YOU MUST ATTEND and got to work.

I ordered the full lineup of *Hot Ones* sauces and they arrived a few days later, nestled like little truth bombs in padded packaging. Tim and I wrote our ten questions carefully; we wanted to dig under the surface pleasantries of "how are you doing?" but did not want to overstep and threaten their sense of privacy. On the appointed day, I bought a huge takeout box of unseasoned wings and two bags of grain-based vegan wing replicas, donned latex gloves and glasses, and tossed the wings in the ten sauces. As the Scoville units mounted, I learned to hold my breath and keep my hands away from my eyes.

The dinner was a surprise, but when we called the boys down and they saw what I'd created—ten wings, each glistening under its own shade of spicy goodness—they smiled in recognition, and our personal episode of *Hot Ones* was under way.

Ten questions, ten wings:

- What could we do differently in our parenting that would make us better parents? (Humble House Ancho & Morita Smokey Tamarind Sauce)

- Which of your friends has influenced you the most and why? (NYC Hot Sauce Verde)

- Ben: Please reveal three heretofore unreported pieces of information about your girlfriend. Finn: Describe the person you are potentially interested in dating, whether or not that person likes you back. (Butterfly Bakery Pink Peppercorn Gin)

- Is there a time when we leaned on our *Gift of Failure* parenting strategy when we should have helped you more? (Heartbeat Hot Sauce)

- Where do you want to go most in the whole world? (Queen Majesty Charcoal Ghost)

- What is your dream job you'd love to have as an adult, if money was no issue and even if it's not a real job? (Dirty Dick's Hot Pepper Sauce with a Tropical Twist)

- What about you reminds you of one or more of your grandparents? (Pepplish Provisions Cranberry Orange Clove)

- What about you reminds you of one or more of your parents? (Bravado Black Garlic Carolina Reaper)

- If the house were on fire and you could walk out with only three things, what would they be (people and animals are safe)? (Dawson's Sichuan Ghost Pepper)

- If there was one thing you could change about yourself, what would it be? (Puckerbutt Reaper Squeezin's)

We finished all the milk in the house around question seven, which is when we switched to vanilla ice cream. Between the jokes and the Scoville-induced weeping, we learned more about our boys at that dinner than we had in years. They gave their answers willingly and honestly, not necessarily because of the spice, but because we'd made an effort to connect on their terms. When we make a genuine effort to get to know our kids for who they are—the highs, the lows, the funny—we are far more likely to earn their trust and respect. When we have their trust and respect, we are better able to impart **knowledge** about the risks of drug and alcohol use, establish clear family **expectations**, and offer practical guidance as to how to put knowledge and expectations into real-life **practice**.

MAKE TIME. Put down the phone, turn off the television, and make time to talk.

BE PRESENT. Physical and emotional presence is far more important than providing answers or solutions.

LISTEN. No, really. It's that simple. Your willingness to listen can be far more important than the things you say, at least when it comes to showing your support, love, and respect for young children and teens.

LISTEN WITHOUT AN AGENDA. Sometimes kids just want to be heard, and don't need parents to fix their problems.

STAYING CONNECTED IS USUALLY MORE IMPORTANT THAN BEING RIGHT. Pause for a moment and consider whether your need to be right is asserting itself.

DON'T TAKE THINGS PERSONALLY. Try to remain emotionally neutral. When kids are upset, they don't want to have to tend to our emotional needs on top of their own. Anger, shame, or guilt will only derail the discussion further, if not close it down completely.

RESPOND WITHOUT JUDGMENT. Be supportive and forgiving while helping them figure out how to do better next time.

COMMUNICATE ON THEIR TERMS. Sometimes kids find it easier to write about difficult subjects. Letters, emails, and texts are valuable, tangible proof of your willingness to connect on their terms rather than yours, and often, the communication itself is more important than the form it takes.

PICK YOUR BATTLES. You don't need to go to war over every minor infraction or hint of trouble, especially if the timing is bad from a mood, fatigue, or hunger perspective. Infrequent, meaningful interventions are more effective than constant nagging.

INVEST IN LONG-TERM CONNECTION. Consistent support and respect are more important than any one single instance of support and respect.

BE THE PARENT. Kids don't want another friend, especially when they are struggling. Parents, even with all their rules and uncool attitudes, can offer stability, consistency, and a reminder about family expectations that's reassuring when life feels out of control.

EMPOWER. Kids want to be independent even as they want our support and guidance. Give them opportunities to be successful, competent, and make their own choices. Sometimes that's going to mean they have to face their own consequences, and that's important for their growth, too. The goal is to raise kids who don't need our help anymore, and as emotionally painful as that can be, it's the most powerful, unselfish gift we can give them.

Tips for Talking about Substance Use

Talking to kids about substance use and other risky behaviors is so difficult. Consider the acrobatics Tim and I have engaged in over the years to make these conversations easier: chairlifts, car trips, hot wings, and careful selection of mood, timing, and snack proximity. If there's a trick to make these moments less torturous, I'm in.

START. While most parents know their kids will have to make decisions about using drugs and alcohol, far too few talk about it. Among parents who do talk about it, they tend to focus on tobacco, alcohol, and marijuana and ignore the risks of other drugs, especially prescription drugs. If you don't talk about it, your kids are at higher risk for use and abuse.

START EARLY. Twenty-nine percent of twelve- to fourteen-year-olds say they have close friends who use substances. That percentage rises to 61 percent by ages fifteen to seventeen. Begin your long-term education in substance use and abuse with general talk about their health and safety, then, as they mature, get into the specifics of substance use beginning with the substances they are most likely to encounter early, such as nicotine and alcohol, and progressing on to the health and legal ramifications of harder drugs as well as the effects of these substances on the brain and mood.

KEEP IT UP. The most effective school- and community-based prevention programs start in elementary school and continue into early adulthood, so your family prevention program should, too. The topic of substance use should be frequently discussed and evolve at an age-appropriate level as your kids grow up.

DON'T FORGET TO LISTEN. Your child's attitudes and intentions regarding substances are powerful predictors of future behavior. If they are favorable toward drugs and alcohol, or talk about their desire to use them in the future, they are significantly more likely to do so than a kid who disapproves of drug and alcohol use.

CREATE OPPORTUNITIES TO TALK. Discussions about healthy behaviors don't just happen on their own. Plan for them, seize opportunities for discussion, and realize that perfect opportunities hardly ever arise. Use the moments you have, when you have them.

TODAY IS ALWAYS PREFERABLE TO LATER. Talking about personal health and safety is hard, whether you are discussing drugs, sex, consent, mental health, self-injury, or a family history of substance use, and it is not going to magically become easier tomorrow, next week, or next year. Talk now.

THE MORE YOU TALK, THE EASIER IT GETS. Frequency normalizes difficult conversations. We talk about substance use and abuse most weeks in our home and those discussions have become as easy as talking about good nutrition.

DON'T WORRY ABOUT BEING A HYPOCRITE. If you drink or use drugs yourself, you can still tell your children not to use. As outlined in chapter 3, drugs and alcohol do more damage to developing brains than to adult brains, some of it permanent.

Special Considerations: When Substance Abuse Runs in the Family

I got sober when my kids were eight and thirteen, and substance abuse has been a regular part of our family conversations ever since. It's had to be; the weight of the research reveals that our best defense as a family is knowledge, clear expectations based on that knowledge, and consistent reinforcement through natural consequences. Years of studies on fraternal and identical twins indicate that about half of a person's risk for substance use disorder depends on genetics and the rest comes down to environmental factors such as ACEs, mental health issues, and the other types of childhood trauma. Given this heightened risk, communication around substance abuse may sound different in a home with a history of substance abuse than one without.

Janelle Hanchett, author of *I'm Just Happy to Be Here: A Memoir of Renegade Mothering*, as well as a recovering alcoholic and drug addict, uses this knowledge to help her children understand their heightened risk. She says she's honest with her kids about her history because they must understand that for them, drinking and substance use is "playing with fire."

She started talking about drug and alcohol use very early on with her children, and now that they are teens in the Netherlands, where the legal drinking age is eighteen, she continues to talk even when their experiments terrify her. When one of her children admitted to trying pot, Hanchett asked about her daughter's experience, why she chose to do it, what it was like, and how she felt after. When her daughter reported feeling depressed afterward, Hanchett was able to capitalize on that observation to explain the chemical reality of substance use: for every high there is a corresponding and ever-deepening low. When her daughter got high, her brain went into chemical overdrive in an effort to adapt and return to a state of equilibrium. "If a drug makes you feel relaxed, withdrawal and craving are experienced as anxiety and tension. If a drug helps you wake up, adaptation includes lethargy; if it reduces pain sensations, suffering will be your lot," writes neuroscientist and recovering addict Judith Grisel in her book, *Never Enough: The Neuroscience and Experience of Addiction*.

If Hanchett had freaked out, screamed and yelled and lectured, they never would have had this discussion, and her daughter never would have gained the knowledge that may keep her from using next time. Even so, Hanchett does not have it all figured out. She still wonders about the impact of her own substance abuse on her kids in between worrying about their use, but adds, "One thing I do know: I have shown my children what a sober life looks like. We have built a raucous, loving home in recovery. It may not be European-cafe glamorous, but it's real, and it's ours, and it will always be my children's no matter where this life takes them. Somehow, that has to be enough."

Substance Abuse Prevention Through the Years

Parenting for prevention starts early and changes along with your child's experience, maturity, and knowledge. If I wanted to give first graders a sense of Charles Dickens's Victorian England I might not read them *Bleak House* during circle time, but I would read Mary Pope Osborne's *Magic Tree House #44: A Ghost Tale for Christmas Time.* The following guidelines are meant to serve as a general guide for easing kids into a discussion about health, safety, and substance use, a conversation that starts in preschool and continues through college.

Preschool and Kindergarten

Knowledge about substance use begins with building healthy behaviors such as washing hands, brushing teeth, eating healthy food, and getting enough exercise. Little kids have short attention spans, so keep explanations short, and don't be offended if their attention drifts. From a cognitive perspective, kids aged four to six love to practice their emerging abilities to count, sort, estimate, and identify numbers and letters, so use that enthusiasm. They learn best through telling or being told stories, movement, rhythms, melodies, and patterns, and there are lots of great ways to incorporate these teaching methods into lessons about healthy choices.

Spell Out Expectations

As you teach kids healthy behaviors, make your family expectations clear. Your expectations will be specific to your family's beliefs and values, of course, but some examples may include:

- We brush our teeth after every meal.

- We wash our hands after we use the bathroom and before meals.

- We eat a healthy diet.

- We don't drink alcohol (or we don't drink alcohol until we are twenty-one, or we drink responsibly; this is up to you, but now is a good time to think about your genetics and personal habits and remember that kids who are raised under the clear expectation that they will not drink until twenty-one are far less likely to develop SUD during their lifetime).

- We don't take medicines we don't need or medicines not prescribed for us.

- We don't take illegal drugs.

- We prioritize sleep and try to get enough of it.

Practice

Putting expectations into practice means reinforcing good decisions with praise and allowing kids to experience consequences for
bad decisions.

Reinforce good decisions with praise as soon as possible after they make that good decision and be specific. "Good job" is
meaningless to a child, but "I'm so proud of you for sticking with
your toothbrushing for two whole minutes!" is meaningful, helpful,
and specific feedback. Feel free to lay on the praise especially thick
when kids make good decisions for their health and well-being in
the face of social or media pressure to act in a way that's contrary
to family expectations or not good for them. For example, if your
expectation is that your child will wash her hands after she uses the
bathroom, and you see her do so when you know she's in a rush
or excited to get out to play with her friends, say, "I know it's hard
sometimes to take a minute to wash your hands when you are excited to go out and play, so I was really proud of you when I saw you
wash your hands after using the bathroom today at Katie's house."

When kids fail to meet expectations, help them figure out ways
they can do better next time. For the child who stayed up too late
and is tired, talk about how you can help support him in getting to
bed earlier tonight. Don't solve the problem for your child, support
them in coming up with answers themselves. Some will be silly,
some will make no sense, but one or two will work, and this is an
important opportunity for your child to not only learn from his
mistake but to feel a sense of self-efficacy about his ability to come
up with solutions for his own problems and put those solutions
into effect.

It's a win-win. You get a happy, alert child and he gets to feel a
huge sense of pride in his own accomplishment.

Use Age-Appropriate Examples and Strategies:

- ESTABLISH AN ENVIRONMENT where kids are allowed and encouraged to ask questions and bring up concerns.

- DON'T LECTURE. Children tune out lectures but are more apt to engage in two-way discussions.

- MAKE SURE THEY HAVE ACCESS to other adults they can trust.

- If you have a relative who is actively abusing addictive substances in the immediate or extended family, PHYSICALLY AND EMOTIONALLY SHIELD YOUNG CHILDREN from the chaos those people can cause. As kids get older, talk about the substances that cause chaotic, destructive behavior. "We don't spend a lot of time with Uncle David because he sometimes drinks too much alcohol, and it's not good for any of us to be around that."

- MAKE UP STORIES ABOUT THE BACTERIA that live between their teeth and turn the sugar from that after-school snack into acids that make cavities in order to talk about why we brush our teeth—that it's not just because we want them to be clean and shiny but because strong, healthy teeth will help us to chew our food into tiny pieces and keep our bodies healthy.

- MAKE SILLY COMMENTS ABOUT USING PRODUCTS such as toothpaste, soap, and shampoo in the wrong way in order to help kids learn that they can make us sick if we use them incorrectly. For example, "What would happen if I put this shampoo in my mouth instead of on my hair?" or "Why don't we just swallow the toothpaste instead of spitting it out?"

- YOUNG CHILDREN CAN BEGIN TO LEARN that over-the-counter medicines and prescription medicines can help us when we are sick, but that we should only take them according to doctors' instructions or when they are prescribed specifically to us.

- USE THE LABEL ON A PRESCRIPTION MEDICATION AS A TEACHING TOOL. If they know their letters or numbers, ask them to find the letters of their name or the first letter of your name on the label. "Wait, before I take this, can you find the first letter of Mommy's name on the label?" is a great way to teach kids not only that you have a name beyond your parent nicknames, but how to spell it. They can also help locate other information they may need to know in case of an emergency, such as the spelling of their last name or the name of their doctor, which is often included on prescription labels.

- HELP THEM UNDERSTAND THE WHY BEHIND THE VITAMINS OR MEDICINES THEY TAKE. For example, if they have a fever, explain that they are taking this pain reliever because their doctor (or other health professional) said it would help make the fever go away. If they take a multivitamin, they can learn that getting enough vitamins usually happens with healthy eating, but sometimes we may need help getting enough in our bodies.

- HELP THEM UNDERSTAND HOW MUCH SLEEP THEY NEED AND WHY IT IS IMPORTANT FOR THEIR HEALTH. Sleep gives their brain an opportunity to store memories of the things they learn and their muscles get a chance to rest. If they don't sleep well, help them link the physical feeling of being tired with the fact that they did not get enough sleep.

- LITTLE KIDS MAY BEGIN TO NOTICE UNHEALTHY BEHAVIORS IN OTHERS, such as smoking, drinking alcohol, eating a lot of sugar, not eating vegetables, or drinking too much soda. When they do, take the opportunity to talk about why people may make unhealthy choices and how your child can make healthy decisions based on what their bodies need.

- CHILDREN AS YOUNG AS THREE ARE ABLE TO IDENTIFY ALCOHOLIC BEVERAGES and differentiate them from other drinks so they need to hear that kids should not drink alcohol because it harms their brains and bodies, and even grown-ups can get sick from drinking too much of it.

Elementary School

Keep reinforcing lessons introduced in preschool, and as kids' understanding of human biology deepens, so, too, can your lessons about healthy behaviors. Attention spans lengthen and they will begin to hear about topics related to health and safety on the news, stories that can begin to feed your discussions. The more we can anchor these discussions in real-world scenarios, the more relevant the lessons will be.

Knowledge

Kids in first through fifth grade have longer attention spans, are becoming better listeners, and are understanding (and predicting) the connection between their actions and natural consequences more reliably. They still pay close attention to parents and still respect parental wisdom and authority, but the older they get the more they will begin to test their independence and the parameters of family expectations. A touchstone memory for me of fourth grade was during our nutrition unit, when my teacher, Ms. Pack, caught us eating dry Jell-O powder out of a box during a class break. She asked why we were eating a box of sugar, especially during a nutrition unit. Our honest answer would have been that it was the exact opposite of what she was teaching. It felt subversive and rebellious and thrilling to be a little bit bad. Our answer was probably something like, "It tastes good," and I'm sure she tore her hair out after dismissal that day. This kind of pushback against authority is typical and healthy as kids move toward their tween years, when they begin to individuate and separate from their parents in earnest.

As they begin to individuate, they will also be exposed to more influences outside the family, including media depictions of substance use. Kids see an average of twenty-three instances of alcohol advertising in the media every month. One study found alcohol played some role in the plot of one out of every eleven cartoons coded in the study and appeared in that cartoon an average of three times. Multiple studies conducted over a few different years found between 52 and 57 percent of popular contemporary G/PG movies depict alcohol use and 30 percent of those instances were branded use of alcohol by an actor. These numbers are significant because alcohol advertising works. A 2006 study found that advertisers spend $6.80 per capita in local markets and for each per capita dollar spent above that, young people drink 3 percent more. And then there's sports advertising. Internationally, alcohol accounts for 20 percent of all sports sponsorships. Those names on your child's favorite soccer player? Those are sponsors' names, and they pay big bucks to get a prime spot on athletes' bodies or on the boards along the side of soccer stadiums and hockey arenas. Anheuser-Busch, for example, spent $350 million in 2016 for the right to display their brand name as prominently as possible in U.S. sports and integrate the experience of drinking their beers with enjoying professional sports.

Early elementary school is a great time to start using what's called "inoculation messaging" to help protect kids from threats to their health and safety. The inoculation theory of communication takes its name from the science of vaccines, which work by introducing a weakened version of a virus in order to confer protection against the real thing. The measles vaccine, for example, is a weakened version of the virus that teaches the human immune system how to mount a defense against the more potent live virus. The inoculation theory of communication as it relates to protecting kids from substance abuse goes a little like this: if we arm kids with a counterargument against reasons they should use alcohol or drugs ("Come on, it's no big deal," or "Come on, everyone does it"), it will shore up their defenses against a real-life, more potent version. Inoculation theory has been shown to be a powerful tool in protecting kids against all kinds of risky behaviors, including smoking, binge drinking, and unprotected sex. In fact, one study found that inoculation messaging can confer so-called cross protection, that messages meant to protect college students against one risky behavior (binge drinking) effectively protected them against other risky behaviors (unprotected sex) not mentioned in the messaging. All of this is a fancy, scientific way to say that rehearsing ways to (forgive me) say no to drugs can not only help kids say no, it can also protect them from engaging in other risky behaviors. The research suggests that rehearsing or inoculation messaging works not just because it gives kids ready access to scripts they can use when they feel at risk or threatened, but because the very existence of a valid counterargument is reassuring to kids. Further, knowing they can use counterarguments increases kids' sense of self-efficacy, and thus their sense of competence even in the face of adversity. I will offer some concrete ways to use the inoculation theory in the practice section, below.

Expectations

Continue to reinforce family expectations around healthy and risky health behaviors while expanding or modifying them as needed. As elementary school–aged children begin to gain more independence and have experiences outside of your family, use them to compare those experiences and influences with your own family's values around health behaviors such as smoking, sex, and substance use. Young children tend to make friends based on proximity and ease of access, while older children begin to seek out friends based on attributes they are curious about, so discussing your family expectations in the context of what they are beginning to see out in the world is an ongoing, constant, and ever-shifting process that will depend on the influences they encounter. Consider the influence of older siblings as your children mature, and remember that the positivity or negativity of an older sibling's peer group affects the kinds of peer groups younger siblings will develop. Continue to provide kids with information about risks associated with substance use, reiterate consistent and clear family expectations on a regular basis, and allow kids to feel the consequences of their unhealthy decisions against a backdrop of conversation, love, respect, and support.

Practice

Here are some ways to put knowledge and expectations into practice for elementary-aged children:

- TAKE TIME TO WATCH WHAT YOUR CHILDREN ARE WATCHING and talk to them about what they are seeing, hearing, and reading.

- USE THOSE EXPOSURES TO ALCOHOL AND DRINKING IN FILMS to talk about fictional characters' behavior. Compare healthy drinking habits to unhealthy ones. You could say, "That character seemed to have a drink every time he was sad. What are some other ways he could have dealt with those feelings?"

- TEACH KIDS TO BE CRITICAL OF ADVERTISING and look for the hidden messages and sales tactics. When they see an ad for beer that portrays young adults as carefree and happy while they drink on a beach with their friends, ask, "What do you think that commercial wants you to buy? Why is it effective or ineffective?" Understanding how the media sells to children and young adults will be a valuable tool for them as they get older and begin to approach the world with a more critical eye.

- TALK ABOUT THEIR EXPERIENCES AT OTHER PEOPLE'S HOUSES in a relaxed and casual (as opposed to a suspicious) way. Ask them if their friend's parents drink or smoke and what do they think about that? What are their friend's parents' expectations around drugs and alcohol? What would happen if their friend Jeffrey drank some beer; what do you think his parents would do? Talking about what other families might do given a health or safety transgression can be a great way to reinforce your own family's expectations around the same behavior.

- TALK ABOUT CHANGING FRIENDSHIPS and what they mean to them. Ask what they like about their friends and what they don't like. Steal our "high, low, funny" dinnertime tradition and apply that to friendships. Ask, "What are the best, worst, and funniest things about your friendship with Jenny?"

• BEGIN TO GIVE KIDS SO-CALLED INOCULATION MESSAGING. Talk through scenarios kids might expect to encounter in the next few years.

What might you say if your friend Amy wants you to go up on the highest monkey bars, where you don't feel safe yet? What if she won't take no for an answer and tells you that you are the only kid who isn't brave enough to use the monkey bars? Talk through valid counterarguments your child can make in this situation and help her feel comfortable making them, that she's likely not the only kid nervous about the monkey bars, and even if she is, it is her right to make decisions about what she does with her body and her safety. Remind her that fear is often our brains and bodies alerting us that something may be bad for us; it's a protective mechanism that's evolved in humans to keep us safe. If she feels scared, it is her right to honor that feeling and reject Amy's pressure.

"What would you do if someone brought a pill into school and offered it to you? What would you say? What if they said it wasn't bad for you?" Offer information on the danger of taking other people's prescriptions, on the harm drugs can do to their developing brain. They could get sick and even die if they take medications intended for someone else.

• START TO TALK ABOUT ESCAPE OPTIONS they can use to get away from risky situations. I have told my children that one of the most important things I do for my sobriety is give myself permission to leave a place where I don't feel safe, even if I'm worried I might offend someone or make them angry. I can always apologize and explain after, but I have never had a friend get mad at me when I explained that I left because I did not feel safe. Real friends want us to be safe and healthy, after all.

- EARLY ELEMENTARY SCHOOL IS THE PERFECT TIME to start talking about what makes for healthy and unhealthy peer relationships. I will go into much greater detail in chapter 8, but begin by asking your child what it is about their relationships that makes them happy. If a relationship seems to be making your child anxious or sad, say, "You know, I notice when you come home from Amy's house you feel sad. Do you know why that is?"

- MODEL AND TALK ABOUT YOUR OWN POSITIVE RELATIONSHIPS. For example, you can talk about how your friendship with Kevin makes you feel confident and happy, and how you decided not to be friends with Ivan anymore because he made you feel uncomfortable and sad.

Middle School

Welcome to the deep end of the pool. Most kids who try addictive substances do so during middle school, so this is no time for polite euphemisms or metaphors. There's a very good chance your kid knows more than you think they know, even if their knowledge is based on YouTube references and sitcom jokes. Don't assume these references are sailing over their heads unheard, as middle school kids hear everything and store it away for future reference. If you talk about these references when they happen, you are more likely to be heard today, before they experiment with substances or other risky behaviors.

Knowledge

Here's where a lifetime of conversation and honest discussion really pays off. Most kids first use substances in middle school, so the tools you will use to prevent substance use and abuse during these years should ideally already be in place. Transitions, especially the move from elementary school to middle school and from middle school to high school, are especially risky periods for kids, so talk and support are vital now. Kids need to know that if they start using alcohol now, they are much, much more likely to develop substance use disorder than if they wait until their brain is done developing in their early twenties, and the earlier kids start using any addictive substance, the more damage it does to their brains and bodies. It can change the way their brain works in both the short and long term. While drugs can make you feel happier or less anxious for a short period of time, they can also decrease the chemicals in your brain that make you feel happy, and consequently, you can end up feeling more sad, anxious, or bored than you were in the first place.

Here is the substance use prevalence data for eighth graders as reported in the National Institute on Drug Abuse Monitoring the Future 2018 survey. This knowledge will be incredibly powerful fodder for discussing ways to think about family expectations, as well as for helping kids develop strong resistance against peer pressure. When they say, "but everybody does it," you can use this data to show them that no, that may be their perception, but the reality is that very few kids their age do it:

2.2 percent of eighth graders used cigarettes in the past month

2.1 percent used smokeless tobacco in the past month

6.1 percent vaped nicotine in the past month

2.6 percent vaped marijuana in the past month

8.1 percent vaped "just flavoring" in the past month

8.2 percent had "more than a sip" of alcohol in the past month

2.1 percent got drunk in the past month

5.6 percent used marijuana or hashish in the past month

1.8 percent used inhalants in the past month

Expectations

Before we get to family expectations, let's talk about some expectations you should have about your child as they enter middle school. First, peer groups begin to eclipse family in importance and influence just as tweens and teens spend more time away from family, are less supervised, and more apt to get bored. Remember, adolescents' dopamine levels are lower in their brains than they were in childhood, so they will feel more bored and require more stimulation and risk to feel normal. Their attraction to risky behaviors is developmentally appropriate, but if you know it's coming, you can plan for it and use it for your benefit.

You can also expect your child will be offered or have opportunities to try addictive substances at some point soon. Given that reality, our job as parents is to keep reiterating and reinforcing family expectations and consequences, and continue to reinforce increasingly important self-advocacy skills and self-efficacy, and make sure they have counterarguments and exit strategies for the challenges to come.

Children raised in families with permissive drug and alcohol expectations, who allow for sips or experimentation in the home, drink earlier, drink significantly more, and experience more negative consequences from their drinking than kids who are raised by parents who consistently communicate clear expectations for drug and alcohol abstinence. Kids who start drinking in middle school have a lifetime risk of substance use dependence of 41 percent. However, if those kids were to wait until eighteen, their risk falls to 17 percent. If they wait until twenty-one, their risk of becoming dependent on alcohol bottoms out at 11 percent.

Middle school is also when kids may begin to ask about whether or not you did (or do) drugs, and how that fits into your family rules. When I asked one expert in adolescent development to give me advice on guidelines for talking to kids about substance use, he advocated for total abstinence until after eighteen. "The advice I give parents is that you should have an unrealistic no alcohol or drug use policy until they are out of high school. Period. And know that your child is going to disobey that. But that doesn't mean that you shouldn't have it that way. When kids know that their parents have a premise of latitude, they're more likely to use."

When I asked him how he'd deal with the inevitable question, "Did you ever try drugs, Dad?" he admitted he may have made some mistakes on that front:

When my son asked me and my wife about our drug and alcohol use when we were kids, I believed that honesty was the best policy. That's something I think I would have handled differently with our son. I was lucky. I don't have the genes for it. I partied hard when I was in the last two years of high school and throughout college. Experimented with just about everything, and I told him that. He told me later, you probably should not have told me that. I said, "I didn't want to lie to you." But he said, "Yeah, but you made it sound like [drugs and alcohol] were kind of cool and fun. If you did it, then why shouldn't I?"

Practice

Here are some ways to put knowledge and expectations into practice for middle school–aged children:

- USE FACTS TO BATTLE PERCEPTION. Knowledge is power when it comes to putting family expectations into practice. If she knows that "everybody does it" is not factually accurate, she can be prepared to counter the most predictable arguments for why she should try it. If she hears, "Come on, everybody does it," she can be prepared with the counterargument, "No they don't. Only two out of every hundred kids do."

- TALK ABOUT CONFUSING MESSAGES IN THE MEDIA. When they see their favorite entertainment or sport celebrities promote drinking or drug use, remind them that drugs and alcohol can cause damage to the brain and body, but they cause much more damage to a young, undeveloped brain. There are plenty of examples to point to in the celebrity world for drug and alcohol use gone bad: Heath Ledger (actor, 28), Amy Winehouse (singer, 27), Tyler Skaggs (baseball pitcher, 27), Whitney Houston (singer, 48), Mac Miller (rapper, 26), Lil Peep (musician, 21), Philip Seymour Hoffman (actor, 46), Prince (musician, 57), Axl Rotten (wrestler, 44), Janis Joplin (singer, 27), José Fernández (baseball pitcher, 24).

Teens

Adolescence may be when kids are most likely to use drugs and alcohol, but if you are just beginning your prevention efforts now, you are going to be doing a lot of makeup remedial work. Your kids have likely already been offered drugs and alcohol; they have seen kids drink or tried it themselves; and they know who they could go to if they wanted to buy stimulants or ecstasy or pot. Whether you are starting now or building on a long history of prevention education, continue to focus on shoring up competence, helping kids make good decisions, encouraging them to set meaningful goals, exploring ways to manage their stress, and building healthy relationships.

Family dinners get very interesting when our family expectations are tested with kids' newly burgeoning skills of argumentation, linear thinking, and logic (their frontal lobes are finally coming online!). You may hear them say things they don't really believe but want to test out by saying them out loud (think divergent religious, political, or cultural stances), but this is good. This means they are trying out identities and ideas and sampling the world around them as they find out what kind of adults they want to be.

Expectations

Adolescence is hard on family expectations. They fray, they get trampled on a little, and that's to be expected as teens test, push, stretch, and experiment with the limits of those expectations. They want—no, need—to know how far our expectations extend, how durable they will be in the face of various situational ethics, and what will happen if they violate them. Given this reality, open communication remains their lifeline. They have to make complicated social and emotional decisions every day, and they are going to make mistakes. If teens feel as if they can trust us to remain supportive and present, even during difficult conversations, we are more likely to hear the truth from them. We can't, as Lisa Damour writes in *Untangled*, parent only for the social options we wish our kids had. When we do, we put our teens in an impossible situation: "they can give up their social lives and stay home with us or they can sneak around and lie to us about what's happening at the parties they're attending." Amid the uncertainty and shifting allegiances, consistent family expectations can be a reassuring ethical and moral touchstone in tonight's decision making, and consistent parental support gives them the freedom to be honest and open tomorrow night at dinner.

Practice

Here are some ways to put knowledge and expectations into practice for high school–aged children:

- KEEP AN EYE ON THE BIG PICTURE. You can diffuse emotionally charged discussions by getting out of the weeds of the specifics and back out in the wide view. If open communication and prevention of future unhealthy behaviors is your goal, does it really matter who said what and when, or who sipped from the stolen alcohol first?

- THINK LONG TERM about your parenting goals.

- EMPATHIZE. You don't have to agree, but you do need to listen, and to do that most effectively, try framing your listening according to adolescent psychologist Lisa Damour's advice: "I've found a tone that communicates that I'm neither critic nor judge; I'm just interested in siding with the teenager's wise, mature side to see if we have any reason to be concerned about her ability to take care of herself."

- USE THEIR GOALS AS LEVERS. The only motivators that really matter to teens—to humans, really—over the long run are intrinsic motivators—the goals, interests, and priorities we set for ourselves. Use them to frame real and hypothetical decisions teens may make. If, for example, they want to be a better runner, smoking will make that more difficult. If they want to move from second chair to first chair in the orchestra, drugs such as pot will mess with their short-term memory and make it more difficult for them to remember that audition piece.

- COMBAT OPINION AND FEELINGS WITH FACTS. If your child claims pot is harmless and helps him relax, combat that argument with the fact that from a brain standpoint, pot is not harmless (and see above, could mess with his goals), and while it can make problems go away for a few hours, they will always be there when he comes back down. In fact, drugs and alcohol tend to compound anxiety over the long term by creating more problems, such as lower grades, short-term memory loss, stress on relationships, and financial drain.

- WHEN THEY ASK ABOUT YOUR DRUG USE, orient your conversation toward the bigger picture. For example, if you have used marijuana and they ask if you have used it, you can admit that you have used it but that you noticed it messed with your short-term memory, or that the pot you used was very different from the pot that's available today. The concentration of THC (tetrahydrocannabinol) has increased substantially over the years, and today's pot can be fifty times stronger than the weed you smoked.

- USE THE MEDIA FOR PROMPTS. If the media is going to portray substance use and report on it in the news, don't avoid it, talk about it. Television shows, your experiences out in the world, news stories—all of these are prompts for discussions. For example, if vaping is in the news, ask your child how people at his school behave. "Do kids at your school vape pot, nicotine, or just flavors?" or "What are your peers more likely to smoke, tobacco or pot?" (This came up recently in our home and I was interested to hear that my sons are more likely to see pot than tobacco.) You could discuss lawsuits against opiate companies ("Did you hear about the settlement with Perdue Pharma? What do you think about opiate manufacturers' responsibility to consumers?"), recent arrests in your neighborhood ("What are people at school saying about that arrest of the drug dealer near the school?"), news of an overdose ("Do you know if your school keeps naloxone on hand for overdoses?").

- TALK ABOUT TEENS' DEVELOPING COGNITIVE STATUS. Helping teens remember that their periodic lapses in organization, planning, and time management are due, in part, to their still-developing brains can help alleviate their stress. Their frontal lobes are not yet complete, and joking about this reality can dispel frustration, anger, and anxiety. We refer to these lapses as "brain farts" and "prefrontal short circuits" when the boys walk into a room and forget why they are there, or fail to pull themselves together in time to get out the door. (This also makes me feel better about my own middle-aged lapses.) If teens understand that their brains are still developing when it comes to executive function, they will also be more likely to understand discussion around addictive substances' heightened danger to those developing minds. Teens can have trouble generalizing ideas from one context to another, so help them get there from time to time.

- LISTEN FOR SMALL CUES. If a teen mentions a health or safety concern in a small way, you can assume it's much bigger in their minds. Throwaway comments are hardly ever meant to be thrown away when it comes to teens. If they trust you enough to offer a reference to drugs or drunk driving or sex, that's a conversational door you should open.

- OFFER OPPORTUNITIES FOR POSITIVE RISK, CONTROL, AND SELF-EFFICACY. Teens crave sensations, risk, and drama because, as mentioned in chapter 3, their brains have lower dopamine levels on a day-to-day basis than little kids or adults. Give them opportunities to find that dopamine they crave or they will make opportunities themselves. I love this line from one of my students, "When I reach the top of the mountain, I feel as if I am on top of the world. This is the defining moment where time slows down and I feel free from all the stress and annoying stuff happening in my life. I can see clouds below me as I look down the steep mountain. I truly feel unstoppable and on top of the world." Encourage kids to break out of their routine and sample experiences that scare them, such as reaching for an ambitious goal, trying something new, auditioning, or taking lessons in a new discipline.

I have so little time alone with my kids these days, and it's getting more difficult to lay claim to more. Between my work and travel schedule, my kids' expanding school, work, and peer obligations, and their increasing need to build lives that have less and less to do with us, it's only going to get more difficult. I plan to make the most of the time I have left, though, and fill it with as many dinners, discussions, and stacks of colorful wooden animals as I can.

CHAPTER 8

EVERYONE'S DOING IT

Friendship, Peer Pressure, and Substance Abuse

My introduction to Brian was in a photograph posted to my son Ben's Facebook timeline. Ben, then fifteen, is in the center of the photo, along with a friend I'd met on his right and a wiry little kid on his left whose name, according to the tag on the photo, was Brian. They are both in their cross-country team warm-ups and Brian is clinging to Ben, his arms and legs wrapped around him like a koala bear.

I probably smiled when I saw that first shot of Brian. Ben had recently transferred to a new high school in Vermont, and I was thrilled to see he was making new friends. At the time, the picture served as evidence that his decision to change schools had been a good one, reassurance that I could stop worrying.

Brian kept showing up in Ben's timeline. At a fall dance, posing between cornstalks and a garland of autumn leaves. In an elevator, making funny faces for the camera with Ben. At the Vermont State Cross Country Championships, stretched out long and lean in a full sprint toward the finish line.

And then he vanished.

Ben explained over dinner in November that Brian had been expelled and shipped off to rehab in Maine. Ben and his friends were upset and worried but steadfast in their support. If Brian did well in rehab, the school had promised, he'd be allowed to return in January. Two months later, he returned to Vermont and from what I could get out of Ben, Brian was doing well—focused, sober, and attending recovery meetings off campus a couple of times a week. I knew these things because I asked. A lot.

As a sober person and teacher of kids with substance use disorder, I was happy for Brian and proud of my son for being supportive.

As Ben's mother, however, I was terrified.

I was proud of Ben's dedication to his friends, his loyalty, and his sense of empathy. Ben had faith in Brian, saw his potential as a student, an athlete, and a human being, and was willing to go to bat for him, willing to spend his time and emotional resources on someone he could have easily cut out of his life. I also knew that Brian needed friends who could be counted on to make healthy choices, and hoped his friends would stick around long enough to give him a lifeline as he learned how to live as a sober adolescent.

On the other hand, I knew what the research says about peers and substance abuse risk. Pick up any article or book on adolescent substance abuse and "peer cohort" is always listed as one of the most, if not the most, significant influences on a teen's decision to use. I'd been teaching adolescents in an inpatient drug and alcohol rehab for a few years at that point, so I also knew how complicated recovery can be for anyone, let alone an adolescent boy living far away from his family. Relapse is a common—even expected—part of adolescent substance abuse recovery, and no matter how proud I was of Ben for his empathy and constancy, I did not want my son to be anywhere near Brian when relapse happened.

And happen, it did.

I wanted to understand more about Brian's story from his perspective, especially about the impact his peers have had on his use, abuse, and recovery. Brian's story highlights a lot of important truths contained in the research on peer influence and substance abuse, but it also highlights many of the problematic, confounding factors inherent in that research.

Brian, who is of Indian and Pakistani descent, was adopted as an infant by his parents, a successful, professional couple of European descent. When I asked Brian to characterize their relationship, Brian replied, "I love my parents and they've done so much for me, but we've never been close. We've never had one of those relationships where we can just talk. Our relationship is pretty surface level and very academic. We talk about schoolwork because that's where we're comfortable." There was not a lot of emotional warmth in their home, he explained, and his parents worked very long hours.

Brian never saw his parents drink alcohol, as they had both opted for sobriety before he was adopted. His father had decided not to drink, Brian said, "because he feared his reactions when he used to drink." Brian's maternal grandmother had been an alcoholic, which had been very distressing for his mother and caused her to avoid alcohol altogether. His parents never explained the reasons for their sobriety, and Brian does not remember hearing about their family history until he was well into his own struggle with addiction.

Brian was at heightened risk for substance abuse well before he entered high school. He'd been seeing a therapist for years to learn how to manage his impulsivity and disruptive attention-seeking behaviors. He had never smoked a cigarette, never had a drink or used a drug, but his behavior was already causing problems. Despite the therapy and his parents' best efforts to help him manage it, Brian was asked to leave his high school in the first semester of his sophomore year. On the advice of an educational consultant, Brian's parents enrolled him in an accredited wilderness therapy program in Utah, a program Brian describes as "a two- to three-month program of initial rehab where they tried to help me learn how to be the genuine me instead of trying to do things and act the way I thought other people wanted me to."

Adolescent drug and alcohol rehabs, and especially wilderness therapy programs, are a notoriously mixed bag of well-trained, dedicated, and responsible professionals who truly care about children, and parasitic opportunists out to make a quick buck by capitalizing on parents' fears and disposable incomes. To paraphrase Wordsworth, when they are good, they can be very good indeed, but when they are bad, they are horrid. Accredited, responsibly managed, evidence-based wilderness therapy programs can be effective interventions for kids who need to learn self-efficacy, resilience, coping skills, executive function, mindfulness, and many other important social-emotional and life skills. There's something about the intense, instructor-led experiences out in the wild that can help kids gain competence and form relationships based on shared adversity and interdependence. However, poorly managed, hard-core, "tough love" boot camp programs have a long record of putting kids' mental health and physical safety at risk.

Brian claims his experience in Utah was mostly positive, and he says he learned a lot about himself over those months. Unfortunately, due to the demographics of the population in Brian's wilderness therapy group, his time in Utah also served as an intensive, three-month peer-taught education in substance use and the mind-set of abuse.

While Brian's instructors were doing their best to teach him how to be self-driven and helping him develop a sense of self-efficacy, Brian was searching for acceptance within a group of teens whose most significant shared connection was their history of substance abuse. Despite the organization's clear rules against drug use war stories, the other members of the group talked constantly about just that. "I was the only one on that team of guys that had never had a drink, had never been exposed to drugs. I didn't even know how one would do drugs or anything, and even though it was pretty clear we were not allowed to glorify that, I was still feeling that I didn't belong. I felt very left out and even though that's something that's okay to be left out of, I'm someone who needed to be part of groups," Brian said.

The research on this topic seems simple at first. Most begin with a sentence like this, from a review article published in the American Psychological Association's monthly periodical, *Psychological Bulletin*: "Peer use of substances has consistently been found to be among the strongest predictors of substance use among youth." Seems simple enough, right? If your kid is around other kids who do drugs, your kid will be more likely to do drugs.

The simplicity of this statement is the stuff of every parent's nightmare: your kid—a good kid, a sober innocent—falls in with the bad kids. They lure him in, offer him drugs, and as a direct result, your good kid becomes a drug addict. This is a tempting cause-and-effect narrative because it allows a parent to view their own kid as a victim and places blame elsewhere, on the so-called bad kids or on the bad kids' parents. *If not for that boy, my daughter would still be sober. If not for the wrong crowd, my kid would be fine today*. This story was exactly what I envisioned when I worried about Ben's exposure to Brian.

Ironically, this narrative is precisely what Brian experienced in Utah, but it's hardly ever how adolescent peer groups form, nor does it accurately describe how peer groups impact a given teen's substance use risk.

The situation Brian faced in his wilderness therapy program was extremely unusual: a socially isolated, drug and alcohol–naive adolescent in the middle of a major life transition was being asked to bond with peers whose common interests and experiences revolved around substance use. If Brian does not learn to assimilate, empathize, and connect with this group, his physical and emotional safety, the future of his education, and his deep, essential need for social belonging will hang in the balance. Literally. The program Brian attended featured climbing, canyoneering, and backpacking in remote, isolated locations.

Now, think back to your peer group in high school and how those groups formed. Were you friends with people who had been selected for you by adults, solidified through isolation, close quarters, and grave danger? Maybe at camp, but hardly ever in the course of everyday life. The calculus of adolescent friendship is not as simple as the cause-and-effect equation parents tend to envision, and adolescent peer relationships hardly ever happen as a result of parental interventions and machinations.

When we are young, our friendships tend to form due to proximity. Our parents were friends, so their kids are our friends. It works out well for everyone; it's one-stop friend shopping. However, as kids move into adolescence, proximity matters less, and kids begin to move out of their childhood peer groups and gravitate toward people and groups they have things in common with; similar personality types, ability levels, identities, or shared interests.

Adolescents also form connections with kids who intrigue them, who possess traits they'd like to possess themselves, or who appeal to their need to experiment with identity and try on new personas. Through this lens, the research on peer influence and substance use gets quite murky.

While it may be true that kids are more likely to use drugs if they hang out with other kids who use drugs, it's also very likely that kids who are more likely to use substances seek each other out in the first place, which sets the stage for higher rates of substance use when they are together. The research bears this out: kids with an interest in substance use or who have been harboring an intent to use are more likely to use. It's also true that most kids' first substance use happens in the company of other kids. There's a little bit of proximity politics at play as well: drug users may become friends because they meet in locations where they use or buy drugs, or because they spend time with teammates playing a sport with higher rates of substance use, such as hockey, football, lacrosse, and wrestling (yes, I checked: cross-country running and other low-contact sports have low rates of substance use while high-contact sports tend to have higher levels). Finally, if a kid has been rejected by friends who don't support his use, the only people left for him to hang out with are the other kids who use substances. On the other hand, kids who want to be sober are more likely to hang out with like-minded kids, either out of a sense of self-preservation or their shared interests, goals, and lifestyle choices.

When Brian graduated from the wilderness program in Utah, he had learned a great deal about drugs and alcohol but had no actual experience using either. He moved to a short-term therapeutic boarding school in Maine in order to finish his sophomore year and allow him to matriculate into a regular school in the fall. For the third time in two years, Brian struggled to transition into a new school and peer group, many of whom were at the therapeutic boarding school to treat their substance abuse. Like the program in Utah, the program in Maine had the potential to be great for Brian. From all appearances, the staff are well trained and know how to help kids like Brian manage their impulsivity, anxiety, and depression, as well as any lingering issues that can arise from adoption, but the nature of Brian's peer group at the school made assimilation and acceptance difficult and, as it turned out, dangerous.

"A lot of the students there had very serious substance abuse problems and were going through relapses. We had very different treatment plans, so this was the second experience of my life where I was out of my element, not really in the same boat as everyone else. Not that I didn't belong there, but I felt different," he recalls.

Brian admits he still craved connection and simultaneously felt as if he did not deserve to be loved, two sentiments that will sound familiar to anyone who has spent even an hour in a recovery meeting or around recovering adolescents. I've never met a kid who really, truly likes himself or feels that he belongs when he enters rehab. That's one thing drugs do well in the short term: they make you feel better about yourself (or at least forget how much you hate yourself) and, when taken with other people who use substances, make you feel a part of something that's larger than your very small, very unworthy self.

Brian finished his transitional program in Maine, and by the start of his junior year, this time at a private boarding school in Vermont, Brian was in desperate need of a literal and metaphorical team. He had been a runner off and on during high school, but he admits he joined the school's cross-country team as much for the social connection as the exercise. "I joined because it was positive, a really good outlet for me, and I loved the social aspect of the team.

We'd go to practice, sit around under the tree to stretch, then go to dinner together. We had a very good core group, and I never really had friends up until then. That's something that was a recurring pattern. Even now, as I have a much better hold of my life, I'm a very social person, I thrive off of positive social energy, but on the flip side of that, it always hits me a lot harder when there's negative social energy," he said.

While the cross-country team provided his daily dose of positive social energy, he had found another group, a loose association of students connected through their shared experiences in wilderness treatment programs, kids who provided Brian with the negative social energy he craved and had not yet learned to resist:

> There were a couple of kids there who had also gone through wilderness treatment, and immediately, there's almost an unspoken connection of like, "Oh, you went through treatment? I did, too. We should talk, hang out." There were two guys, seniors, and they were very much into smoking and drinking, and they met another guy that had also gone through wilderness, so I kind of flocked to be with them because I'm very socially shy until I get to know someone. I internalize a lot and it's really hard for me to become extroverted. So I clung to them, and one of the first things on the first weekends is they said, "Oh, does anybody smoke cigarettes? Do you all want to go down to the tree and go smoke?" And I had never smoked, but I had been exposed to all of that in wilderness and then in rehab, so immediately, I was like, "Oh, I want to be part of this group."

In an effort to earn his place in this group, Brian played up his experience with substances and his involvement in rehab programs.

"I was able to talk about all the crazy stuff that I had witnessed and been around. I wanted to play it off like I wasn't a novice to this because I wanted to be accepted, and I think because of that, it did snowball very quickly because they thought, 'Oh, he's already kind of experienced with this.'" In order to catch reality up with the exaggerations and lies he'd been weaving, Brian's substance use quickly escalated from cigarettes to alcohol, then pot, and on to mushrooms, acid, and K2 (also known as synthetic marijuana, even though it bears no chemical relationship to marijuana). In retrospect, Brian says, he had problems with substance abuse, but the substances were not his true addiction. If he had to identify his drugs of choice, they would be risk and acceptance.

> It just sort of spiraled. I kept telling myself, it's not a physical addiction of needing to use these substances, it's the social aspect. And I still truly believe that was most of it. There was definitely an underlying addiction and need to use the substances, but a lot of it was that I was addicted to these feelings of being with people and doing this rebellious, high-risk, not-okay activity. I was addicted to chasing these feelings of belonging.

While the research on peer influence and substance use is complicated, Temple University psychologist Laurence Steinberg has a clear and elegant explanation for Brian's attraction to "being with people and doing this rebellious, high-risk, not-okay activity."

Steinberg wanted to understand the impact of peer groups on risky decision making in adolescence, so he recruited three groups of people, half male, half female, from three different age groups (13–16, 18–22, and over 24). He had them all play a computer game called "Chicken," in which participants have to make decisions about whether or not to stop a car they control once a traffic light turns from green to yellow. The objective of the game is to get as far as possible without running into the wall that appears when the light turns red. The beauty of this game is that it measures the risks study participants take in real time rather than the risks they might take in a hypothetical situation. For the purposes of the study, the participants played the game alone, alongside same-age peers, and with peers waiting outside the door while the participants played the game inside by themselves.

According to the study, adolescent participants took significantly more risks when other adolescents watched them play. Their peers were not cheering, not egging them on to go further, go faster, to run that yellow light—they were just there, in the room, watching quietly. The adults, on the other hand, took the same amount of risk, no matter whether they were alone or with peers. "For reasons not yet understood the presence of peers makes adolescents and youth, but not adults, more likely to take risks and more likely to make risky decisions," Steinberg concludes.

This finding holds even if the researcher lies to the subject and there are no peers watching. As long as kids believe they are being watched, they will engage in riskier behaviors. This is the best explanation I've heard for the insane stunts adolescents pull for the sake of YouTube views. The very idea that another kid may be watching them, on some laptop somewhere, is enough to egg them into lighting their farts on fire, eating inadvisable amounts of cinnamon, or engaging in something my younger son described as "Bum fights, but believe me, you don't want to know." I made the mistake of googling it, and he was right. I didn't.

Brian, like most adolescents, was willing to engage in more risk and break school rules when the kids from the wilderness programs were around, even if taking those risks meant he could lose just about everything else he'd come to value. Brian's behavior makes sense not only in the context of Steinberg's research on peers and risk, but when analyzed through the lens of adolescent brain development. Brian knew he'd get kicked out if he broke school rules and got caught; after all, he had direct, firsthand evidence of that particular cause-and-effect equation. An adult might be able to weigh the risk of expulsion from a second high school in two years against the momentary thrill of smoking an illicit joint and decide on the prudent course, but teens calculate risk differently. It's not that they don't understand the possible consequences of their actions, it's that they value the positive benefit (forging social connections and winning social approval) more than the negative risks (expulsion, loss of his Vermont friends, teacher and parental disapproval, to name just a few). It can be helpful for us to understand how adolescents calculate risk not only because it gives us an answer to the question, "What on earth were you thinking?" but because if we understand their upside-heavy calculations, we can help them adjust their math.

Despite the looming risk of those negative consequences, Brian started blowing off cross-country practices to smoke and drink in the woods. Then, while drinking vodka in the woods one Sunday afternoon, Brian noticed that one of the girls in the group was acting the way he used to act with his substance-using friends, playing up her experience and drinking more than she could handle, and she quickly progressed past drunk to alcohol poisoning. "She was not okay. Not able to walk, not able to talk. It was terrifying. And at that point, everyone we were hanging out with ran. They left. It was just me, her, and her friend, and I didn't want to leave, I was not okay leaving, so we had to call the ambulance, obviously. She was peeing herself and throwing up, unable to control anything," Brian recalled.

Once the ambulance arrived and he was sure she was in capable hands, Brian left to check in with his dorm parents. For the next twelve hours Brian had no idea if she was alive or dead, and he had good reason to believe she might be dead. He had been extremely drunk and high when the ambulance arrived, but he believes he heard them say something about not being able to find her pulse. He had given the paramedics his name, so he was not surprised when the police arrived on campus the next day to question him. Shortly thereafter, Brian was expelled from school but offered a reprieve: if he did well at the rehab program (at the therapeutic boarding school he'd attended in Maine the year before) he'd be allowed to return to school in January.

Brian did just that. He excelled in rehab and returned to school in Vermont in January with a renewed sense of purpose. He was sober, stayed away from the kids he used to drink and smoke with, and focused on school, running, and attending twelve-step meetings off campus. He finished the year strong and healthy, but once he was back home in North Carolina for the summer, he started hanging out with a kid who had also attended wilderness camp. To no one's surprise, they spent the summer smoking weed, taking mushrooms and LSD, and using fake IDs to get alcohol. "Even though," Brian added with a laugh, "I looked eleven years old. It was absurd. But, there it was again, that need to be connected, and I had no friends."

Brian returned to Vermont for his senior year and tried to stay focused on running, academics, and his positive group of friends, but he also made friends with a group of seniors who liked to party. "They didn't have the serious addictive personality and my history of substance abuse, and they were able to have house parties and just drink casually and just have a fun time," he said. Unlike his friends, Brian was not able to drink casually and just have a fun time. On Halloween, the day he ran in the Vermont State Cross Country Championships and qualified for the New England Championships, Brian was expelled for the second and final time for attending a house party where alcohol was being served.

And with that, Brian was done. After all the transgressions, expulsions, and other consequences, this was, in the language of recovery, his turning point.

> Something clicked in my head. I threw away so much. The best moments of my life were at that school in Vermont. Athletically, academically, and socially, and I really needed to be able to not depend on substances to have good relationships with people. I was being a selfish brat and it wasn't just affecting me. Mr. R and Ms. L called me out on it, too. They said, "You're being so inconsiderate to your friends. Think about how sad you are right now because you're losing all of them, they're all also losing you." [My friends] would be worrying about me when they should be worrying about themselves and having fun, and being able to focus on schoolwork and athletics. . . . I was so sick of letting people down and negatively affecting them because all I ever say is, "My friends kept me so positive and they did so much for me." But I needed to do things for them, too.

This last observation, right here, is what the research on peer influence does not often control for: the circumstances in which peers teach each other not just how to become better and more self-aware, but also better able to empathize and advocate for each other. One journal article phrases the effect this way: "even though adolescents typically initiate and use drugs when they are with their peers and even though drug-using adolescents often affiliate with drug using friends and groups, peers do not unilaterally provide a context that supports substance use. In many cases peers may even be more likely to encourage *nonuse* [emphasis theirs] than to encourage substance use. The vast majority of adolescents report that their friends would *not* condone frequent drug use or experimenting with illicit drugs other than marijuana."

This effect comes into play in evidence-based, proven substance abuse prevention curricula like the education programs covered in depth in chapter 9. When parents, schools, and communities strengthen kids' sense of empathy, social responsibility, and self-efficacy, and nourish children's interpersonal skills, prevention programs are a lot more likely to be effective. Just as families can establish and enforce family norms, peer groups (as well as schools, communities, and other organized social groups) can assert and enforce their norms for the betterment of all.

I used the power of indirect peer pressure to my advantage all the time in my classroom, and it was often much more effective than active enforcement on my part. Here's an example of a time when it worked particularly well: The kids in my rehab classroom were talking about triggers that made them crave drugs. One kid mentioned that a piece of student art someone had hung up in the back of the room looked a lot like the LSD blotter paper he used to eat, and we agreed as a class to take the poster down so he did not have to suffer through cravings while we were supposed to be focused on a writing assignment. The student began to talk a little too romantically about his LSD use, so I interrupted him and changed the subject to a topic I'd been excited to share with them anyway. The student became incredibly angry with me for interrupting, and even after I apologized, he refused to talk to me. About an hour later, he interrupted me while I was explaining something else. A student to my right said, "Aren't you going to get angry at him for interrupting you, just like he did?"

"No," I said. "He got excited about what he wanted to tell me. That happens sometimes."

I cast a meaningful glance and smile over at the student to drive my point home.

The class came down on him like a ton of bricks, but because I'd smiled, everyone was in on the joke. The rebukes were firm, if softened by humor, and the student apologized for interrupting me.

I established a healthy norm, a student defied that norm, and the group enforced it for me.

The secret sauce of parenting and educating for substance abuse prevention is to help our kids form and maintain positive, healthy relationships, harness the positive social pressure in their peer groups, and equip all kids with the skills they need to stay safe and healthy, no matter who they hang out with.

Here are some practical ways to do just that.

Tips for Helping Kids Build Healthy Peer Groups

- BE RESPONSIVE TO YOUR CHILD'S EMOTIONAL NEEDS. Kids' first lessons in attachment happen in infancy, when they learn whether their caregivers will be responsive to their emotional and physical needs. These first lessons in social attachment carry over into their adult relationships and shape their expectations. If no one responded to their needs as an infant, they are not likely to expect their peers or romantic partners to do so, either.

- BE YOUR CHILD'S EMOTIONAL ROLE MODEL. Kids learn what healthy friendships look like from us. If we maintain relationships with people who hurt, demean, undermine, and sabotage us, our children are more likely to form the same kind of relationships.

- ASK YOUR KIDS WHAT THEIR IDEAL RELATIONSHIPS WOULD LOOK LIKE. Think of this as goal-setting for friendships. If your child knows what they want in a friend, they will be more likely to befriend those kinds of people.

- TALK TO YOUR KIDS ABOUT THEIR ACTUAL RELATIONSHIPS. Listen, observe, and comment. If one of your child's friends makes her feel bad about herself, or if a friend is excessively competitive, deceptive, or mean, talk about how that makes your child feel. Similarly, if your child has a friend who is kind, generous, or goes out of his way to raise up others, talk about that. Years ago, one of my son's friends defended a kid who was being bullied, and you'd better believe that was the topic at dinner that night and many dinners after. We still talk about that kid's bravery and empathy, a decade on.

- TAKE STOCK OF YOUR OWN FRIENDSHIPS. If you have friends or romantic partners who treat you badly, your child is learning to have the same kind of expectations in their own relationships. Model healthy self-respect by cutting those people loose and teach your child to value positive relationships over negative ones.

- MODEL HEALTHY FRIENDSHIPS. Think about the things you value in your relationships and talk about them. If your friend noticed you were sad and texted to make sure you were okay, talk about that as a kind, empathetic caring thing to do. If someone treats you badly, and you stand up for yourself or end that relationship, talk about that, too.

- KNOW THE PARENTS OF YOUR CHILDREN'S FRIENDS. As neurologist Frances Jensen writes, "you are really sharing parenting with all the parents of your kids' friends," and she's absolutely right. It's one of the reasons Tim and I drive four hours roundtrip to give Finn time with his two best friends in our former hometown in New Hampshire. Their parents are part of our village—people I trust to make sensible, healthy decisions when they are around my child.

- HELP KIDS BUILD A STRONG SENSE OF IDENTITY. When kids know who they are and what they stand for, they are much less likely to be swayed by peer pressure, and the best way to help children and teens build identity is by respecting, supporting, and encouraging their individuality.

Tips for Helping Kids Resist Peer Pressure

- "NO, THANKS" IS AN UNDERRATED AND EFFECTIVE ANSWER. Most of the time, fewer people care why you do or do not do something. I tend to forget this and get all flustered, thinking I need an elaborate excuse when this often works beautifully.

- NO, EVERYBODY ISN'T DOING IT. It may seem like everybody drinks, or smokes, or has sex, but we (and especially adolescents) wildly overestimate the rates of risky behaviors in their peers. Unfortunately, when kids (or groups of kids, or communities) believe more people are using drugs and alcohol, they are more likely to adapt to that perceived standard and use drugs and alcohol themselves. Make sure your kids have accurate information about the prevalence of risky behaviors so they don't rely on their misperceptions.

- VOLUNTEER TO BE THE DESIGNATED DRIVER. Your friends won't be mad at you for not drinking; they will be thrilled they have a sober ride home.

- HAVE A SECRET WORD OR TEXT to use with friends or parents to indicate you'd like to leave. It could be an agreed-upon word, a facial expression, an emoji, whatever allows you to let your friends or parents know that you don't want to stay and need a ride home.

- TELL PEOPLE ADDICTION RUNS IN YOUR FAMILY and you'd rather not drink. This excuse worked for my older son for years, but it does not have to be true to use it. Feel free to throw a hypothetical relative under the bus and cite genetics as your get-out-of-drinking-free card.

- TELL PEOPLE YOU ARE ALLERGIC TO ALCOHOL. Intolerance to alcohol is a real thing, and it's due to a genetic condition. Some people can't break the alcohol down in their body, and it causes hives, trouble breathing, low blood pressure, diarrhea, nausea, and vomiting. The cure for alcohol intolerance? Not drinking!

- TELL PEOPLE YOU'RE ALLERGIC TO AN INGREDIENT IN THE DRINK. Gluten intolerant? Sorry, can't drink most beer. Wine intolerant? That's a thing, too, affecting a little over 7 percent of people in one German survey.

- SAY ALCOHOL GIVES YOU MIGRAINES. According to the American Migraine Foundation, alcohol is a migraine trigger, and as a result, people prone to migraine headaches drink less than the general population.

- TELL PEOPLE YOUR PARENTS DRUG TEST. As long as we are throwing relatives under the bus, toss your parents under there, too. Your peers might think your parents are helicoptering jerks, but that's fine. We don't mind. Your safety and sobriety are more important.

- SAY YOU ARE ON A MEDICATION THAT DOES NOT MIX WELL WITH ALCOHOL. There are many common medications that specifically say right on the label that you can't drink alcohol, either because it makes the medication less potent or compounds the effects of the alcohol. I'd go with the first reason because it sounds serious, and it's true of antibiotics as well as a lot of other over-the-counter and prescription medications.

- SAY YOU HAVE A SLEEP DISORDER. According to the National Sleep Foundation, alcohol is terrible for sleep. It interrupts your circadian rhythms, rhythms that are already off due to something called an adolescent sleep phase delay. It also blocks REM sleep, the deepest, most restorative sleep cycle, makes breathing disorders such as snoring and sleep apnea (pauses in breathing patterns) worse, and generally interrupts your sleep both for brain-chemistry reasons and having-to-pee reasons (alcohol is a diuretic, so it makes you have to pee more often). Even if you are a healthy sleeper, alcohol wreaks havoc, so citing a sleep disorder such as insomnia or apnea is a great excuse.

- HOLD A NON-ALCOHOLIC DRINK if you want something in your hand to blend in with the crowd, or take a beer and empty it in the bathroom sink so you can refill it with water if you don't want to be the only one with no drink in your hand. It works, but be careful. School counselors advise kids not to hold drinks or colored plastic cups at parties because when colleges or future employers look into social media, those cups will automatically imply underage alcohol consumption.

- SAY YOU ARE WATCHING YOUR WEIGHT. I'm not a fan of dieting, but if it means you get out of having to drink, go for it. Cite an upcoming wrestling match, track meet, school dance, modeling shoot, whatever. Alcohol contains a lot of empty calories, after all, and when most people start trying to lose weight their doctors often tell them to cut out alcohol as an easy first step.

- KNOW THAT YOU CAN LEAVE. Sometimes, when we get caught up in the energy and momentum of a crowd, we tend to forget that leaving is an option. I worry people will think I'm rude or lame or old and boring. Since I got sober, however, leaving has become a reasonable response even when I'm pretty sure people will think I'm rude or lame or old and boring. Give kids a reliable exit strategy. Make sure they know they can call at any time, with no repercussions, guilt, or pressure to tell on other kids if they need a way out of a situation that makes them uncomfortable. This promise can be a part of your family contract (see chapter 6).

Advice for Kids Who Are Worried about a Friend's Substance Use

- SAY SOMETHING. Just bringing up your concern is the hardest part, but it's also the most important part. It lets the friend know they are seen, and even if they don't want to talk right at that moment, it lets them know they have a sympathetic ear when they do want to talk.

- BE THE FIRST PIECE OF THE PUZZLE. I often use a puzzle metaphor when talking to kids about substance abuse. Hardly anyone stops drinking or using drugs when the first person mentions there might be a problem, and it may not be until the fiftieth person says something. But without person one through forty-nine, there is no fiftieth person.

- BE PATIENT. Admitting that you can't control your substance use is incredibly difficult. When I finally said it after a decade of thinking it, I threw up. Keep that in mind when talking to a friend about their substance use.

- LISTEN. You may not hear the words you want to hear, such as "I think I have a problem with alcohol," but you might hear them talk about feeling out of control, or lost, or disconnected. These are all ways of expressing the powerlessness they may feel when it comes to their drinking or drugging.

- DON'T BE AFRAID TO ASK FOR HELP. If your friend drinks too much or takes too much of a drug, call 911 and stay for the ambulance. One of the bravest things Brian did was stay to make sure that girl with alcohol poisoning got help. Yes, there were consequences for Brian because he stayed, but those consequences not only got him closer to help, they ensured that the girl lived, and she very well could have died that night.

- REAL LIFE PLAYS OUT MORE SLOWLY THAN MOVIES AND TELEVI-SION. Your friend may not admit to her problems and head off to rehab, where she thrives and finds herself in time for the commercial break. Stay positive, and be patient. This is a miniseries, at least. Maybe even a long-running drama series.

- WE CAN'T FIX OUR FRIENDS. If that were true, Brian would have been fixed after his first or second expulsion. We can't live our friends' lives for them. I love the line in the serenity prayer in which we ask for help, "to accept the things I cannot change, the courage to change the things I can, and the wisdom to know the difference." I can be pretty bad at that, especially when it comes to my students and my kids. I really want to change their lives for them, make their lives better, but sometimes I can't and that eats me up inside. This line reminds me that it's okay to let go of other people's stuff and focus on carrying my own load.

- TALK TO AN ADULT. In chapter 9, I tell the story of a girl named Georgia who was a full-blown, daily drinking alcoholic by the time I became her teacher in high school and her friends came to me to figure out what to do. We could not save her—she had a lot of years of using drugs and alcohol ahead of her before she figured out how to save herself—but those girls told me it felt good to unburden themselves. They had been keeping Georgia's addictions a secret for so long it had become a habit, and they were exhausted. Today, a sober and healthy Georgia is grateful to these friends for telling, and wishes someone had told even earlier. Teachers, school counselors or nurses, a doctor, someone at your church—all of these people have been trained in what to do when kids are having problems with drugs or alcohol, and will do whatever they can to help.

Brian spent the remainder of his senior year in recovery and salvage mode at another boarding school in Maine. He focused on academics and met every behavioral expectation his parents, teachers, and school administrators set for him: he did not smoke; he did not drink; he did not even break curfew. And through it all, he kept running. Today, Brian is in college, working on a dual degree in psychology and business, and he has big goals: "Someday I would like to open up my own rehab program and try to use approaches that nurture the whole person, realizing that everyone is different, everyone belongs in a place and shouldn't need to compromise who they are to be liked. Less focus on substance treatment alone and more emphasis on well-being, because everyone can benefit from that."

Brian and Ben are still friends, and for all my fretting and maternal worry, I'm grateful for their relationship. Brian has benefited from the support of his friends, but I'd argue that they have benefited from knowing and supporting him, too. They had a front-row seat to the drama as it played out before them. They watched as their school set high expectations for his behavior and held him accountable when he screwed up. They were there for him on his last morning in Vermont, a story Brian told me at the very end of our phone conversation.

> I still have dreams about the morning I had to leave Vermont for the last time. Ben and some of my other friends woke up early, drove to school or left their dorms to meet me, and we were allowed to go for a last run together. We went for our normal loop run and I realized right then and there that they supported me more than anyone had through all of that. I could not be where I am now without those relationships.

Brian's Facebook photos are very different today than they were when I first spotted him clinging to my kid. I recently scrolled through his albums, comparing the adolescent he was to the man he's become. He's still attention-seeking, still hamming it up for the camera, but in my favorite photo, he's uncharacteristically still. He's being lifted into the air by four friends on a beach. One woman supports his feet, two men spot him on either side, and another woman stands in front of him, her arms outstretched in case he falls. He's focusing on some fixed spot in the distance as his friends hold him aloft, his arms raised high into a clear blue summer sky.

CHAPTER 9

THE ABC'S OF ADDICTION PREVENTION
Best Practices for Schools

The students in my Salt Lake City private high school class-room looked pretty much like what you'd expect students on the wealthy end of the valley to look like in that long-ago fall of 2001. Happy, healthy teenagers geographically and emotionally insulated from the world beyond their neighborhoods, planning where they would ski that weekend and what time they should meet at the coffee shop down the block. Our huge, airy classroom faced southward, overlooking the tidy grid of downtown and beyond, past Point of the Mountain and on to the Great Salt Lake. Behind us towered the Wasatch Range, with its snowy peaks and comforting constancy. The Olympics were coming to town, and the city shone especially bright that fall. The graffiti, the litter, even the city's homeless had been carefully hidden away from the eyes of the world. The view from our classroom that autumn was hopeful and bright.

For one girl in my class, however, that fall marked the beginning of a dark descent that would consume what was left of her childhood. Georgia looked like her classmates, and could sometimes function like them, but while they skied and drank lattes, Georgia struggled to survive her overwhelming anxiety and depression with vast quantities of alcohol.

I knew she was sad and hurting—anyone who spent even an hour with her could see that—but I had no idea how far she'd already descended into her own personal hell when I became her teacher. By the time she dropped out of high school and moved in with her drug-dealing boyfriend, she'd been lost to us for a while.

Saving her at that point was nearly impossible, but in the years since I watched her walk out of my classroom for the last time, I've wondered what we could have done differently, what conversations, what education, what interventions might have prevented the decade of trauma, pain, and self-medication she only barely managed to survive.

Georgia and I spent many hours on Skype, texts, and email exploring every aspect of her journey, from her earliest memories of an anxious childhood to her life as she understands it now: a woman who has made a lot of mistakes but who has also worked hard to find stability, love, and peace.

Georgia grew up in an affluent neighborhood in Salt Lake City with her mother, father, and older brother. Mental illness and substance abuse ran in her family, but even though they had lost multiple relatives to suicide and overdose, she said, "I don't think I ever remember us talking about [substance abuse] until I started getting in trouble for it."

Georgia, née Mary Anne, was named after a paternal aunt who committed suicide before Georgia was born. While she never met her aunt, being Mary Anne's namesake carried a lot of significance for Georgia and, in retrospect, shaped part of her identity. "Everybody in my family says I'm exactly like her. She was fiercely independent and rebellious and wanted to go into theater, and did a lot of drugs, a lot of partying. From what I hear from my relatives, it sounds like she was probably an addict and an alcoholic," Georgia said.

By the time she was in elementary school, Georgia was already uncomfortable in her own skin, a descriptor I have heard a lot from my students and other people in recovery. Georgia's emotional discomfort manifested as anxiety and gastrointestinal symptoms. "I was a super anxious child. I remember being in third grade and getting these horrible stomachaches, and I thought I was going to die. I would make the school office call my mom, and she would pick me up and take me to the ER, but they were like, 'There's really nothing wrong with her. She's fine. Physically, she's fine.'"

Mentally, however, Georgia was anything but fine. She wasn't even within shouting distance of fine. No one thought to offer mental health counseling, though, so Georgia did her best to cope with the stomachaches and panic attacks on her own. "I got better at acclimating to my anxiety, but it was still there. Then when I hit about eighth grade it got worse. Middle school is so harsh; the social dynamics, everybody wants to be cool. Everybody wants to fit in and feel like they have a place. I didn't, and I just remember it being really brutal."

By eighth grade, Georgia had run out of coping mechanisms and was desperate for a way to escape her constant anxiety, fear, and sadness. And then, during a school drug and alcohol abuse prevention assembly, she found it. Georgia's middle school did what so many schools do in their efforts to cobble together an addiction prevention program: they invited a recovering alcoholic to speak about his experience.

Georgia recalls, "He seemed impossibly old, and I think he was doing the best that he could, but where I was in my mind, all I heard was 'There's drugs and there's alcohol and they make you feel numb, make you not feel things.' I just ignored the rest of the story, all the horrible consequences that he experienced."

Georgia didn't start drinking because a friend pressured her to try it, or because she wanted to get a weekend buzz on. Georgia started to drink because she was looking for relief from herself and a way out from under the constant, crushing weight of her anxiety. Her first taste of that relief came in the form of Wild Turkey, pilfered from her parents' alcohol supply. For the first time in years, she said, she felt normal. "I felt relieved. I was like, 'This is amazing, and I want to feel like this all the time, because if I can have this all the time, I would be okay.'" Georgia started stealing from her parents' liquor cabinet and taking alcohol to school in a water bottle.

I asked if anyone noticed or mentioned her drinking to her.

"Yeah. So my mom eventually noticed. I think she didn't say anything for a while . . . I don't know. I think it's sort of like it went against etiquette or something. You were expected to just behave a certain way and not talk about your feelings or things that were upsetting for you."

By the time I became her English teacher in junior year, Georgia had been a daily drinker for at least two years.

Life got more challenging for Georgia when her parents went through an acrimonious divorce. Her father, who had long been distant and emotionally removed, moved out and became an even more elusive presence in her life. As the consequences of Georgia's drinking began to pile up, however, she began to shift her allegiance over to her father. Her mother, who held her accountable and set limits, was getting in the way of Georgia's drinking and the chasm between them grew wider as Georgia's dependence on alcohol deepened. "There was no way to have a conversation with [my mother] about my drinking, because she could not understand. She thought I could just stop, and I felt I needed it to survive. What does my old boss always say? 'The opposite of addiction is not sobriety, but connection'?" Whether you agree with Georgia's old boss or neuroscientist and author Judith Grisel, who writes in her book, *Never Enough*, "The opposite of addiction [. . .] is not sobriety but choice," Georgia had neither. She'd drifted so far away from her family and the concept of sobriety that she no longer felt as if she had any choice but to keep drinking.

Georgia isolated herself from everyone who cared about her and tried to get between her and the booze. Her relationship with her mother had deteriorated so badly that Georgia moved in with a boyfriend two years her senior. Her father allowed her to stay at his apartment from time to time, but she did not have much of a relationship with either parent by the end of her junior year. From Georgia's perspective, she'd found the perfect situation. "[My boyfriend] had his own apartment. He had a car. He had drugs. He had an ID that said he was twenty-one, so I was like, 'That's my ticket.'"

She may have scored a useful boyfriend, but Georgia's true love was alcohol. It had become her constant companion, the one thing she could depend upon, and the answer to all her problems. Even her friends Polly, Danielle, and Kaitlin were beginning to lose hope and patience. They stuck around until the demands of keeping her present in her own life became too much. By the fall of her senior year, her friends were at their wits' end. Her friend Polly recalls, "That fall Kaitlin and I would show up to Georgia's boyfriend's house every morning before school with clean clothes and a sack lunch and drag her to school." Eventually, it became too stressful to keep propping Georgia up, so her friends let her go out of self-preservation. "I don't think I realized how traumatic losing a friend to addiction was at the time. I grieved her like she had actually died . . . and moved on with my life," Polly recalls.

Looking back, Georgia admits she did not realize how rapidly her life was falling apart. She could not comprehend the consequences we could all see barreling toward her, and even if she had, quitting was never an option. Even as she lost her family, friends, and education, she says, "None of that seemed as urgent as the fact that I just felt so awful inside all the time. I needed [alcohol] to survive. It was the only thing I was really interested in." But alcohol had long since stopped providing any relief. According to the self-perpetuating cycle familiar to alcoholics, the more she drank, the sadder and more desperate she became. The sadder and more desperate she became, the more she needed to drink.

Georgia never really decided to quit high school, she recalls. She simply woke up one morning and realized that somewhere along the way, she had. "I don't even remember making that choice. I just remember drinking heavily and doing a lot of drugs one night, and becoming aware again about a month later. I think somebody else told me that I couldn't go back. I wanted to go back, but I couldn't. I don't even remember quitting."

I've spent almost twenty years wondering what we could have done to save Georgia from what I knew was going to be a very bleak future. If we'd only had a different substance abuse prevention program, if only we'd made this referral or that recommendation. If only I'd called her parents, checked in with her friends, asked about her sadness. Kids spend eight hours a day, 180 days a year in school;

surely we have the time, resources, and expertise to offer kids like
Georgia an addiction prevention education that works?

In fact, we do. Unfortunately, very few schools are using it.

The Evolution of School Health and Prevention Programs

Modern school health programs began in 1850 when professionals
in the emerging field of public health realized schools could play an
important role in promoting community health by inspecting and
vaccinating children enrolled in public schools. It worked; vacci-
nation rates went up and absenteeism went down. Fifty years later,
when Lina Rogers, the first school nurse, was hired to work in the
New York City schools, administrators further improved community health by referring students out for medical care and teaching
them and their families about good hygiene. Today, research shows
when schools have a full-time school nurse on staff, attendance
for poor and otherwise marginalized students improves, students
with chronic illnesses get more regular care, and immunization rates
go up.

Among their many other public health and prevention duties,
modern school nurses are often responsible for teaching health
class, long understood as a euphemism for sex education. For years,
sex ed consisted of little more than cartoons of sperm swimming
toward an egg in the context of abstinence-only messaging. However, the AIDS epidemic of the eighties and nineties forced schools
to reconsider this simplistic approach and widen the scope to include information on preventing sexually transmitted diseases, usually in the guise of lessons devoted to condoms and bananas. More
recently, the #MeToo movement launched a national discussion
on consent, gender, and power in sexual relationships, and health
curricula broadened once again to educate kids about these issues.

Just as the definition of sex ed broadened, so, too, did the definition of student health. Personal hygiene and vaccinations are still a part of most school health curricula, but today, "health" implies a holistic approach to promoting positive self-concept, personal safety, and healthy behaviors in kids while allowing teachers and administrators to screen for students at risk of physical, mental, or emotional problems.

Substance abuse programs have long been understood to play a major role in the promotion of healthy behaviors, yet a mere 57 percent of high schools implement any of these prevention programs at all. Of those 57 percent, only 10 percent are based on any sort of evidence that the program works.

The substance abuse education Georgia experienced in middle school was standard for the eighties and early nineties. Education programs focused on scare tactics like Nancy Reagan's "Just Say No," the Drug Abuse Resistance Education (DARE) program, and the "This Is Your Brain on Drugs" ad campaign. In the 1980s, DARE was the preeminent model of prevention in schools, but it was based on a legal approach to substance abuse rather than a developmental or cognitive approach, and focused on the lowest possible expectations for human behavior. These efforts, while well intentioned, did not work and in some instances were counterproductive. According to studies conducted to assess the effectiveness of DARE, students who had participated in DARE courses were more likely to make unhealthy choices around drugs and alcohol than kids who had not participated. In other words, a DARE education was worse than no education at all. Many of these programs don't even begin until middle school, and for many kids, that's too late.

Shortly after DARE was shown to be counterproductive, the federal government decreed that in order to be funded, substance abuse programs must be evidence based, and several registries have emerged to guide schools in their selection process. The thread running through all successful programs is their focus on overall health as a starting point for any discussion about substance abuse. Kids do not become substance abusers in a vacuum, and if we want to prevent childhood addiction, it's vital we look at every child's eco-system. Effective programs strengthen kids' emotional and social skills and give them tools for resisting peer and cultural pressure to use addictive substances while establishing a communal norm that prioritizes healthy habits and decision making.

The good news is that many schools are already implementing these sorts of programs in the guise of SEL, or social-emotional learning programs. The acronym was coined in 1994 as an umbrella term for a whole range of concepts, competencies, and skills, and has come to be defined as "the process through which children and adults acquire and effectively apply the knowledge, attitudes, and skills necessary to understand and manage emotions, set and achieve positive goals, feel and show empathy for others, establish and maintain positive relationships, and make responsible decisions."

I can't help but read that definition in the context of Georgia's story and imagine how different her life could have been if she'd had her own personal army of teachers, counselors, staff, and coaches all trained in SEL education, focused on helping her build these skills in the classroom, in individual conversations and conferences, on the sidelines, and over conversations at lunch.

When I wasn't thinking of Georgia, I imagined my students in the rehab: the boy who moved from foster home to group home to foster home for his entire young life and never learned how to name his emotions or talk about how they influenced his actions. No one ever talked to him about setting goals. I know, because I taught him and his classmates two entire classes on goal-setting, and he told me it had never occurred to him—not once—to set a big, ambitious goal and then formulate and write down the short-term, achievable goals he'd have to plan and execute to get there. I thought about the toxic relationships my students form with friends who shame and romantic partners who demean. I think of all the decisions those kids must make when they leave the safe confines of rehab and return to their homes, schools, and communities, and feel like a wartime doctor in a M*A*S*H unit. Detox 'em, patch up their visible wounds, and ship 'em back to the front. If we continue to send kids into the world with bandages over their gaping wounds, lacking the basic emotional and cognitive skills they need to survive, we will continue to lose them to the wages of substance abuse, violence, teen pregnancy, and suicide.

Good prevention programs change lives, and the very best save them. One such program is LifeSkills Training (LST), a program created by Gilbert J. Botvin of Weill Cornell Medical College. LST is a school-based program that has been shown through over thirty peer-reviewed studies to prevent tobacco, alcohol, and drug use, risky driving, high-risk sexual behaviors, juvenile violence, and delinquency. It is being used in schools in all fifty states and thirty-five other countries, and has been extensively evaluated for efficacy and outcomes. While early evaluations of LifeSkills focused on White, middle-class children, later assessments widened to include a variety of ethnicities and economic levels, in both rural and urban settings, for kids in elementary, middle, and high school. LST has been shown to cut drug use anywhere from 50 to 80 percent, prevent initiation of smoking, and reduce alcohol and marijuana use, and the program is equally effective when taught by a variety of instructors, usually in small group settings.

Lecture-based substance abuse prevention education, like the assembly Georgia attended in middle school, is the most ineffective way to teach kids about addictive substances or most other topics, for that matter. What works, especially in substance use prevention programs, is active learning in its various forms: peer-led discussions, small group activities, and role-playing and lessons that extend outside the classroom, into the halls, onto the playing fields, and accompany kids on the bus when they leave for the day. Well-designed and implemented substance abuse prevention programs strengthen bonds between home and school, students and staff. They foster a supportive and safe environment for students where relationships can flourish, and when students feel connected to their schools, researchers have found declines in risk-taking and negative behavior, including substance use, unplanned pregnancy, violence, depression, suicide, and substance use, and students have higher levels of emotional well-being. SEL programs also promote healthy socialization and the formation of cultural norms and beliefs, an aspect of education so critical to learning and development that it's been called the "hidden curriculum." This hidden curriculum is more than a class here, or a lesson there; it's what teachers, coaches, counselors, and administrators teach kids all day, every day through their own behavior and through clear expectations and natural consequences.

I'm often asked by teachers and school counselors how they can implement more effective prevention programs in their own schools, but the honest answer is that one teacher or counselor can't change school culture on their own, and believe me, most of us have tried at one time or another. While Georgia's high school did not offer a formal, structured SEL program (very few did in the late nineties), she credits a few teachers and her high school counselor with keeping her in school and as safe and sane as possible during that time. "The adults I trusted communicated with me as another human being instead of in that punitive 'I'm up here, and you are down there' kind of way. I really appreciated that. I wanted to stay in school, I really did, but by then, I think I was too sick," Georgia said.

Today's professional school counselors possess master's degrees in school counseling and receive specific training in prevention science as a part of that education. Many school counselors also have teaching training and experience, and in some states, counselors are also required to have at least two years of teaching experience in order to become licensed as school counselors. School counselors are ideally positioned to help shape school prevention programs and offer students and their families support around substance use and abuse, as their professional organization, the American School Counselor Association (ASCA), has defined a national model to support counselors as they research, implement, and assess the efficacy of these programs. School counselors are trained to understand that individual, community, and cultural factors influence drug and alcohol use and abuse so it is essential to address these factors and promote relationships between home and school.

In order to be effective, school-based prevention programs need the support of dedicated principals, headmasters, and superintendents who are willing to lead by example and help shape school culture. Yes, school boards and superintendents secure funding, set district or school expectations, implement assessments, and deliver feedback, but even more important, they set the tone through their actions, priorities, and enthusiasm. Staff and students are skilled at sniffing out insincere lip service to mandated programs, or shell programs that look good in a school recruiting brochure, and will respond in kind. However, when school leaders are invested with their heads, hearts, and actions, staff and students are much more likely to be invested as well.

Quality, evidence-based prevention programs can be expensive to implement, but they pay off. Economic analysis reveals that every dollar invested in evidence-based substance abuse prevention programs can save communities as much as $38 in the long run because they won't need to pay for treatment and the crime caused by substance abuse.

I have reviewed many substance abuse prevention programs and life skills training programs, and as I read about their conceptual frameworks and completed sample lessons, my mind kept wandering back to Georgia. She could have benefited so much from the knowledge and skills these programs teach. She could have learned how to manage her stomachaches by mitigating her stress. She could have learned how to talk about her feelings, ask for help, and make better choices. She could have known her worth and valued her life. She could have operated from a place of knowledge about the pros and cons of substance use. She could have had a safe space to talk about the issues her parents never broached and learned how to form and maintain positive relationships in which she was valued.

One evening, while completing a sample exercise designed to strengthen family communication, empathy, and listening skills, I got a little weepy. After calling my parents to check in and tell them I love them and appreciate all their support, I realized my sadness was for Georgia's family, not mine. I was mourning for everything Georgia and her family lost. They have come a long way together, but so much was lost along the way. How different their family could have been if Georgia and her parents had had access to a meaningful prevention program, one that started early, around the time of Georgia's first anxiety-induced stomachache.

Preschool and Kindergarten

That's right, preschool. Effective substance abuse prevention programs really should start this early. Remember, substance abuse prevention is really about the promotion of healthy habits and identification and intervention for social-emotional delays, and that has to start early.

The transition to school is an exciting and difficult time for young children, and it is also when early issues around socialization first emerge. Here's where we start having to say things to teachers like, "Huh, that's weird, he doesn't bite other kids when he's at home," because of course he doesn't. Unless home is an unsafe place to be, kids usually feel most calm, well adjusted, and emotionally balanced at home, but toss that kid into a room with twenty other kids, away from parents and the family dog, and they react in new and surprising ways. Social and cognitive pressures are often the match that can ignite emotional explosions, but this isn't always a bad thing. Helping kids learn to deal with these outbursts in a way that helps them and the people around them is an important part of early education, and the earlier parents, teachers, and school staff are able to intervene and help get kids back on a path toward healthy social-emotional learning, the better.

Positive home-school relationships are at the center of these interventions. When parents and teachers communicate well, just about everything goes more smoothly, from school adjustment to academic performance and the formation of positive peer relationships.

Preschool programs are a balancing act between social-emotional and academic learning, and recent research implies that the proper balance lies on the social-emotional side, especially if we are interested in preventing substance use. Effective prevention programs emphasize the early formation of prosocial behaviors, defined as behaviors that benefit the entire community, such as sharing toys, helping another kid up when he falls, and empathizing with other kids when they are sad.

Kids with strong prosocial traits not only do better in school and life; they will be more likely to improve the lives of the people they come in contact with. Prosocial traits such as empathy, perspective-taking, self-regulation, collaboration, and kindness can be taught, and when taught well, they reduce substance use risk factors such as early childhood aggression. Michele Borba, author of *Unselfie: Why Empathetic Kids Succeed in Our All-About-Me World*, cites a large study of one such program called Roots of Empathy that resulted in "an 88 percent drop in 'proactive aggression' (the cold-hearted use of aggression to get what you want)." Coldhearted aggression in preschool can be managed and remediated, but without early intervention, it can turn into something much more destructive.

Teach and promote prosocial behaviors and you reduce behaviors that contribute to risk. Reduce risk behaviors and you reduce the chances kids will use addictive substances.

Effective preschool prevention and social-emotional learning programs emphasize:

- Clear expectations for classroom behavior.

- Clear natural consequences for violating classroom norms.

- Small group, cooperative learning.

- Supporting kids' abilities to control their emotions.

- Teacher training for classroom management skills.

- Self-care for teachers so they can be positive role models.

As I reviewed the excellent preschool programs available, I envisioned a very young Georgia learning how to identify her emotions, maybe pointing to a picture of an anxious-looking cartoon face on a mood identification card and telling her teacher, "That means worried." One program, Promoting Alternative Thinking Strategies (PATHS), teaches preschoolers to define feelings and emotions, and describe how people who are experiencing a given emotion look like, sound like, and act. For example, if the feeling is "proud," the teacher leads the children through a discussion of what "proud" means. Is it a comfortable or uncomfortable feeling? When have you felt proud of yourself? Can you show others what it looks like when you are proud of yourself? How do you tell someone else you are proud of them?

Being able to identify emotions may sound self-evident, but for young children not yet fluent in the social cues around various emotions, these lessons can help them understand what they are feeling and, equally important, what other kids may be feeling. For kids less able to understand social cues, lessons like this can help them learn how to "read" other people, an essential skill that can prevent social ostracism and early aggression, both significant risk factors in all kinds of adverse outcomes, including substance abuse. Finally, PATHS encourages parental involvement and supplies lessons to bring parents into the conversation to help teachers deal with sensitive emotional topics, such as fear, sadness, or anxiety over conflicts at home. When children have an emotional vocabulary, and school is a safe space to express and identify emotions, kids are much more likely to receive the interventions they need before emotions such as worry, anger, and sadness turn into stomachaches, hitting, and depression.

Elementary School

As children move from preschool to elementary school, substance abuse prevention programs continue to promote prosocial skills and healthy decision making. As young children begin to encounter more stress and conflict in their lives through experience and the media, elementary prevention programs should also include lessons on stress management and mindfulness, social conflict management and communication skills, and self-advocacy training.

Parents and schools continue to build on home-school relationships, one of the most important pieces in the prevention puzzle. When kids feel safe and cared for at school, they are less likely to engage in all kinds of risky behaviors, including substance use and early sexual activity. School connectedness is especially vital for highly mobile kids, such as the children of military parents, and homeless and foster kids, as educational continuity and stability has been shown to be one of the most important predictors of well-being and life success. In fact, one analysis of adolescent risk factors and long-term outcomes found school connectedness to be the only education-related variable that protected kids against every single adverse life outcome. **The more accepted, cared for, safe, and connected kids feel in their schools, the better their lives will be according to just about any measure.**

Around third or fourth grade, discussions about the existence of smoking and vaping should start to happen, including discussion about why people smoke and how it changes their bodies. LifeSkills Training, for example, includes a lesson on the heart, exercise, and taking one's pulse so kids begin to understand how a healthy heart functions. As kids move on to fifth and sixth grades, these discussions include information on types of tobacco products, smoking and vaping laws, and how health research informs legislation on its sale and distribution.

Effective elementary school prevention programs focus on:

- Early academic success and competence, especially around literacy.

- Continuing support for children's sense of competence, optimism, assertiveness, and ability to empathize with others.

- Strong communication skills.

- The formation of good study habits and organizational skills.

- Building good social skills and talking about what makes for healthy relationships.

- Formulating and practicing ways to say no around unhealthy or risky behaviors, including in hypothetical scenarios.

- Shaping community norms around physical health and anti-drug attitudes.

- Discussion about advertising, especially child-targeted advertising and what tools advertisers use to persuade us to purchase their products.

- Helping parents reinforce the learning kids are doing at school around healthy behaviors and anti-drug attitudes.

Georgia would have benefited so much from a substance use prevention program heavy on social-emotional learning skills such as stress management, decision making, and social skills. Given the destructive role her romantic relationships would play in just a few years, she could have benefited from one LifeSkills Training lesson that asks students to think about the purpose of friendship and consider why they make and keep friends. Another exercise asks the student to write down "What makes someone your friend?" and "What makes someone not your friend?" Finally, it prompts kids to consider what kind of friend they are, and provides a list of attributes to choose from, such as caring, honest, reliable, trustworthy, understanding, and funny, as well as a set of blank areas kids can fill in with descriptors they choose themselves. Georgia had good friends like Polly and Danielle, friends who tried to look out for her and act in her best interest, but she rejected those relationships in favor of romantic attachments with people who exploited her. Being able to identify and talk openly about what she needed, and conversely found unhealthy in a relationship, might have helped her steer clear of some of the exploitative relationships she embraced in order to feel needed and wanted.

Middle School

Here's when educators, counselors, and kids really roll up their sleeves and get to work. At this point in their development, many kids are contemplating substance use, and some have begun to experiment already. Remember, according to surveys on adolescent substance use, the earlier adolescents initiate their drug and alcohol use, the more likely they are to progress to substance abuse during their lifetimes.

Middle school, then, is where prevention meets intervention and if schools wait until middle school to start their substance abuse programs, they are running seriously late to the prevention game. Effective middle school prevention programs continue to help kids shore up their social and communication skills, help them resolve social conflicts, manage their emotions, and maintain a healthy self-concept. However, they increasingly emphasize decision-making and refusal skills and help kids understand that humans often overestimate how much their peers drink and use drugs. Educators use facts about substance use rates to counter kids' ignorance and misperceptions, while educating them on the significant risks involved in using drugs and alcohol.

Parents can reinforce this education at home. A popular topic in many substance abuse prevention programs that has become a regular topic of conversation in our family is "What is this ad trying to sell me and how?" In our home, the game often devolves into derision for advertising companies, but it gives the boys an opportunity to think critically about how they are being targeted. Increasingly, we've started talking about how a particular advertiser reached us in the first place. Our online buying habits? Our Google searches? Did we click on an ad for something related? This discussion can also serve as a great jumping-off point for talks about online privacy and safety.

Effective middle school prevention programs focus on:

- Identifying emotions and thoughts and talking about appropriate responses to them.

- Teaching self-analysis so kids can find a more forgiving way to respond to their bad impulses and learn from their mistakes.

- Learning to identify what causes stress and practicing skills such as mindfulness or reframing in order to minimize the physical and emotional effects of that stress.

- Encouraging a regular practice of setting big goals and short-term, achievable goals as a way to stay focused on future success and happiness.

- Talking about ways to reinforce feelings of self-efficacy, agency, and hope.

- Conflict resolution skills that include anger management, negotiation, compromise, and reframing conflict as constructive debate.

- Communication skills, verbal and nonverbal, that help kids avoid misunderstandings and build consensus.

- Effective listening skills, especially when communicating with people who hold different opinions or who have had different experiences from our own.

- Practicing refusing substances and other unsafe situations in pairs and groups.

- Decision-making skills, including discussion about why rules exist, how following them can help kids achieve their goals, and what to do when rules run counter to individual ethics or morals.

- Stress management.

- Anger management.

Middle school was when Georgia's early childhood anxiety and pain collided with opportunities for self-medication, and it was also when she could have used a lot more practical support from teachers and other adults in her life. Unfortunately, Georgia did not know how to ask for help, did not possess the vocabulary to explain what she needed help for, and had few tools for making healthy decisions. As I completed an exercise from LST on healthy decision making, I realized while the lesson on "The Three C's of Effective Decision-Making: Clarify, Consider, Choose" would have helped Georgia make good decisions, she was already lacking so many of the social and emotional tools she needed in order to gather the evidence to inform her choices.

During one of our interviews, Georgia told me a story about how she coped with a few eighth-grade boys who had treated her badly throughout most of middle school. These were kids who had made her feel worthless, who had mocked and taunted her. She hated them, but she also craved their acceptance and an end to the bullying. In her mind, Georgia's best possible solution was to make them like her by agreeing to get drunk with them in a park near her home. Unsurprisingly, her solution did not change their behavior, but it did solidify her belief that alcohol was a powerful antidote to her emotional pain. "I felt like I could talk to these boys who I really wanted to like me. I'd been picked on by them, and I cared what they thought. I was in middle school. You care when you are in middle school."

LST's Three C's may not have stopped Georgia from drinking that night, but they would have given her the tools to make a whole lot of other, better decisions throughout her adolescence and young adulthood. The process outlined in the LST decision-making exercises helps kids clarify the problem at hand, consider a list of possible choices and the natural consequences of those actions, then choose the best possible solution to the problem. Georgia admitted that by the start of high school, she felt so limited in her choices that she could not have created an entire list of possible solutions, let alone consequences. Given so few choices, alcohol very quickly went from being one option in a list of choices to the only possible solution she could imagine.

High School

At this point, substance abuse prevention programs, and SEL programs more generally, are focused on reinforcing ideas and skills around self-management, self-awareness, relationship skills, and responsible decision making so that kids will be able to successfully navigate the transition into college, the workforce, and adulthood. These transitions are scary. The stakes are higher, the consequences for making poor decisions get more and more painful, and stress increases. High school is also the last opportunity teachers and parents have to prepare kids for the many choices they will have to make in college around drug and alcohol use and other risky behaviors.

Effective high school prevention programs focus on:

- Knowledge that combats myths around substance use, including rates of adolescent drug use, its impact on mental and physical health, and adverse consequences such as academic failure, social conflict, and increased life stress.

- Good decision making around healthy behaviors and personal health choices.

- Personal safety and the right to have physical and emotional boundaries.

- Personal risk factors for substance use and abuse, including genetic and environmental risk, personality type, peer influence, and exposure through high-risk sports, with an eye toward the decisions they will have to make in college.

- Continued analysis of media influences, including alcohol and drug marketing.

- Building on a repertoire of tools for managing kids' stress and anxiety, stressors that can cause teens to seek out relief through drugs and alcohol.

- Practicing and improving social skills and cues, including listening, greeting, and social exchanges from small talk to deep, meaningful discussions on challenging topics.

- Building on conflict resolution skills and learning how to diffuse high emotion in challenging discussion and debate using anger management skills, mediation, and dispute resolution.

- How personal needs and desires intersect with school, community, and societal standards, and discovering ways to balance them with the betterment of the greater community.

- Strengthening refusal skills, reinforcing concepts of personal safety and boundaries in the face of peer pressure.

- Assertiveness and self-advocacy practice as well as discussion of how to respond to pressure in the form of counterarguments.

- Healthy relationships: what they look like, how to achieve them, what kind of communication skills to use, and how to resolve conflicts.

- Planning long- and short-term goals and time management.

By the time Georgia entered high school, she was already drinking on a daily basis, cutting herself to relieve her emotional pain, and struggling to get to school every day. By her own admission, she'd given up—on herself, on any hope for a better future, on a way out. "There wasn't anything else that I was connected to. I wasn't connected to my schooling or my studies or to the future, like, 'I want to be this when I grow up,' or 'I want to achieve this,' or 'I'm connected to my peers because we have this bond.' I felt completely disconnected," Georgia recalls.

While it's unlikely even the most effective SEL or substance abuse prevention program could have changed Georgia's trajectory at this late stage in her decline, she still might have benefited from individual lessons around decision making, stress management, and family communication. The high school LST curriculum features an exercise I would have loved to see Georgia complete before she moved out of her mother's home, one designed to promote communication, empathy, listening, and reflection skills by way of a structured interview with a parent or guardian. The lesson is carefully designed to support kids as they improve communication with a family member and has the added benefit of allowing kids to understand why parents behave the way they do and have the expectations they have for their kids. As I read through the lesson, I imagined Georgia asking her mother questions such as, "What did you like to do when you were my age?" and "What rules and expectations did your family have?" because I knew from meeting with her on multiple occasions that her mother loved her daughter despite her frustration, sadness, and disappointment over the direction Georgia's life had taken. Any opportunity for Georgia and her mother to build empathy for each other could have changed the way they related to each other, and I can't help but wonder what their relationship might look like today if they had been offered convenient, helpful, and structured opportunities to connect.

Alas, they did not, and once Georgia quit high school, her downward trajectory steepened. As is often the case with kids who start drinking early in adolescence in order to mask or avoid their pain, Georgia's emotional maturity had not kept up with her biological age. Kids are often emotionally arrested at the age they started abusing drugs and alcohol because they never faced their problems or learned how to manage their lives. "I had no coping skills. I'd spent the last few years by using every drug I could get my hands on. That was the only thing I knew how to do."

By her nineteenth birthday, there was very little left of Georgia, and we fell out of touch. I thought of her often and relied on word of mouth and Facebook to know whether she was alive or dead. According to Georgia, those years were pretty grim.

> I had no purpose in life. I was empty. I had no hope at all. I remember feeling really betrayed by the alcohol and the drugs, feeling like, "I gave up my whole life for you, and you aren't even working anymore." That was the most dangerous place I have ever been. I've been there twice in my life, and it's awful because it's like I can either die, I can kill myself, or I don't know. I didn't know there was such a thing as recovery. At least not for me. I attempted suicide probably once a week.

Despite her many suicide attempts, Georgia survived. Alcohol was no longer her refuge, and as so often happens with substance use disorder, drugs and alcohol work until they don't anymore. When alcohol stopped working for Georgia, she felt abandoned.

Georgia awoke from an alcoholic blackout on the morning of her nineteenth birthday in the University of Utah Medical Center. When her mother called the hospital with a final offer of rehab, Georgia accepted.

> It was just as simple as that. I was just resigned. I was really just waiting for somebody to be like, "Look, this is what's happening. You have two choices." It was just good timing.

I was at a place where I was very willing for the first time in a long time to listen to, "You can either choose to keep doing this and burn your life down, or you can choose this other path. What do you want to choose?"

Georgia thrived in rehab. "I loved it because I didn't feel like crap. The withdrawals, and the hell I'd put my body through, feeling horrible all the time, it was such a prison. When I finally detoxed fully, long enough to start feeling like a human again, it was the first time I felt, I guess I'll call it freedom."

I'd love to report that Georgia has been continuously sober since that day, enjoying the life she always wanted, but that's not always how sobriety works, particularly for young adults. One night, after five and a half years sober, Georgia relapsed with a beer she purchased on her way home from a twelve-step recovery meeting. She tried to explain her thinking at the time: "I thought, I can just stop at the 7-Eleven, and I can just buy a six-pack and see what happens."

What happened was this: as bad as her first go-around with substance abuse was, the second was worse. Just after she picked up drinking again, her older brother, who had been struggling with mental illness, committed suicide. The loss of her brother caused Georgia and her family immense and unmanageable pain, and her addiction to alcohol expanded to whatever would ease that pain. She self-medicated with Klonopin, then prescription pain pills, and ultimately, heroin. She lost her job; her car was repossessed; she became homeless, stole from her parents, moved in with an abusive drug and arms dealer, and then, at twenty-six, Georgia found out she was pregnant.

When she gave birth to her son, she believed her love for him would be enough to keep her clean. "I loved my son so much. He was my whole world. I thought he would be the thing that was going to help me get better," she recalls. However, after months of trying to care for an infant while struggling to feed her addiction, even Georgia had to admit her son was not safe. She did the most sane thing she could think to do: she handed her son over to her family. "It was like putting the baby in the basket and pushing it downriver," she recalls. "I can't go be with my family, but I know that they'll take my son, and he'll be safe. I know I can't leave this situation, but he can."

Once she lost her son, Georgia entered the final stages of her substance abuse. She was homeless, sick, and emaciated, living in a park she could see from the windows of our high school classroom ten years before. The same day Georgia discovered papers had been filed to terminate her parental rights, an old friend traveled across the country to check in on her. He said he'd had a feeling she was in bad shape and that he was supposed to show up and help, so he did. They rode a Greyhound bus to his mother's house in Plains, Georgia, and with the help of his extended family and a local hospital, Georgia got sober.

Three months later, on June 30, 2015, Georgia's parental rights were terminated and her son was adopted into a loving family. I asked her how she managed to stay sober through that, noting that lots of people would have given up. "I knew what it would look like if I just went out and destroyed myself again. Then my son would never know. My story would just be tragic, it would just be this sad thing that happened. Then I would be a mystery to him, and he would never know the truth, how much I loved him."

I keep two photos of Georgia paper-clipped inside the green folder I use to store her interview transcripts, a copy of her brother's obituary, and some handwritten notes. The first photo is from our 2002 school yearbook. Her hair is styled in perfect spiral curls pulled back from her face with barrettes. She's surrounded by photos of her smiling classmates, and while Georgia has a smile on her face, too, hers is just for show; there's no life, let alone joy, in her eyes. The second was taken seventeen years later, on May 7, 2019, at her college graduation. She's wearing her cap and gown, her father on one side of her, her mother on the other. Maybe the photo was taken at an odd moment, but her parents look frankly shocked, not quite sure what to do with this day, so long in coming. Georgia, however, clearly feels no such ambivalence. Her smile is massive and genuine; it starts in her eyes and radiates through her entire body.

Even in the face of Georgia's recovery, success, and happiness, I have so many regrets. I regret I did not see the scope of her problem; she was drunk, right there in class as I prattled on about metaphor, Toni Morrison, and Walt Whitman. I regret I did not understand how desperate and sad she was, and how unsafe her life had become. She was struggling to show up for her life every single day and I had no idea.

A few days ago, I noticed she'd changed her Facebook page to include a quote, the last line from Albert Camus's essay "The Myth of Sisyphus," based on the Greek myth where the title character is sentenced to roll an enormous boulder up a hill only to have it roll back down to the bottom over and over again. Camus's essay contemplates whether suicide is a valid response, given a life as tragic and absurd as that of Sisyphus. No, he concludes, the proper response to a life of struggle is defiance: "*La lutte elle-même vers les sommets suffit à remplir un cœur d'homme. Il faut imaginer Sisyphe heureux,*" or, in English, "The struggle itself toward the heights is enough to fill a man's heart. One must imagine Sisyphus happy."

No matter how deep or sincere my regrets, I can't go back and eliminate her years of struggle, absurdity, and loss, so I follow her directive: I imagine Georgia happy.

CHAPTER 10

HEALTHILY EVER AFTER

Preventing Addiction in College and Beyond

In my second year of college, I became a residence adviser (RA) in Butterfield dorm at the University of Massachusetts Amherst, mainly because I believed it would be a great deal. I had a large single room on the first floor of a beautiful small dorm at the top of a hill overlooking the rest of the campus, my housing was paid for, and I received a check for $25.88 every Friday afternoon. In exchange for living in a dorm best known for its rampant drug use, I was the face of law and order on a floor of students who were nearly all older than I was at the time. I was on call nearly 24/7 for everything from roommate disputes to clogged toilets and medical emergencies. Freshman year I had learned that my role as a peer drug and alcohol counselor was not the way to win a popularity contest but my social status really tanked once I became a nineteen-year-old RA with a side gig as a drug and alcohol counselor.

Despite the high narc factor and the havoc it wrought on my social life, I loved being an RA, and I really loved being an RA in Butterfield. Yes, Butterfield had a long-standing reputation as a drug-infested den of iniquity, but it was also quiet, insular, and quaint. Butterfield was the only dorm on campus with its own in-house kitchen and dining room and, thanks to the Phish-listening, tie-dye-wearing residents, it was also known for vegetarian meals.

Sure, things got a little nuts, but that's what made Butterfield Butterfield. The fourth-floor residents liked to turn their hall into a steam bath complete with Slip-n-Slide, and our beloved dining hall served as a front for a thriving drug market that was later exposed when the university noticed Butterfield's dining hall made more money than all other student businesses combined. And it's true, the dorm seceded once (not just from the university but from the state of Massachusetts and the United States of America), and even sent then-president George H. W. Bush a letter explaining their reasons (the First Gulf War). They even offered the United States and Canada space in the dorm basement for their embassies.

Whatever. The bread was baked fresh daily, from scratch, and the entire dorm sat down for family-style dinners. Those dinners, followed by nightly Ultimate Frisbee matches in the traffic circle out front, bonded us together and I could not have loved my college family any more.

Even so, my college family got up to some shenanigans, and I was tasked with keeping some semblance of order. Once, the second-floor RA called to inform me that two students were very high, very naked, and having sex in the hall outside his door. What, he concluded, was I going to do about it? I reminded him that my responsibilities that night were limited to the first floor, so he'd have to be the one to do something about it.

And that was on a weeknight.

Now, on the weekends it got crazy. One night as I sat in my room studying for a microbiology exam, one of the residents on my floor came down to find me. Her friend, whom I will call Christina, was so drunk they were seriously worried. Could I come down and check her out? I headed down to the other end of the hall and indeed, she was plastered. Blotto. She was semiconscious on one of the beds, and her speech wasn't just slurred, it was incomprehensible. After some asking around, we heard she'd been at a party with house-made punch, and had downed many cups. I could have surmised that on my own, as her lips bore the telltale evidence of Cherry Kool-Aid FD&C Red No. 40.

Her breathing was shallow, so I quickly decided to head over to health services down a steep road and across a campus covered in a sheet of ice. A male resident from down the hall drove us over in his truck. We made it down the hill intact, slid into the emergency entrance, unloaded my now-unconscious dead-weight resident, and carried her inside. As a nurse loaded her on the gurney, she vomited all over me while simultaneously losing control of both her bladder and bowels.

She (and I) survived her bout with alcohol poisoning and she learned a valuable lesson about the nature of binge drinking. I found Christina on social media and messaged to ask if that night at health services changed her relationship with alcohol, and she wrote back within seconds, "Sure. That was quite an eye-opening day. That is when I learned to stay away from hard alcohol. I use it as a reminder when I feel wine while cooking dinner is becoming a habit. Those multiple large drinks were enough to cause a three-day hangover. I really thought I was dying."

She was right to be scared, as she very well could have died that night. Each year, emergency rooms see nearly 120,000 children and adolescents who have become injured or ill due to drinking alcohol, and 4,300 of them die as a result of alcohol poisoning.

Danger in Numbers

Developmentally, breaking rules, drinking to excess, and doing exactly the opposite of what parents want kids to do makes perfect sense. Adolescence, and especially the tail end of it, when young adults are living away from home yet still spend summer in their childhood bedrooms, when they are still financially dependent on parents (and no matter how little they want to admit it, a little emotionally dependent, too), is confusing. In the college years, individuation, the process of pulling away from parents and becoming their own person, is push-and-pull. While most older adolescents test the boundaries of their independence and ability to handle risks away from home, it's especially important for them to have a realistic understanding of who is drinking and how much.

Where college drinking is concerned, perception drives reality. Sociologists call this phenomenon "pluralistic ignorance," and it's a powerful force on college campuses. Despite the ubiquity of the beer-funneling, shot-pounding, barely conscious college student stereotype depicted in movies and television, this experience is far from the norm. However, our societal belief in this stereotype fuels expectations around college drinking habits, undermines enforcement of underage drinking laws, and increases the amount of alcohol consumed on college campuses.

The authors of my favorite academic paper on the topic, "Pluralistic Ignorance and Alcohol Use on Campus: Some Consequences of Misperceiving the Social Norm," conducted experiments at

Princeton University that reveal just how powerful pluralistic ignorance around substance use can be.

Alcohol is (or was around the time the experiments were conducted, in the late eighties and early nineties) important to the social scene at Princeton for both enrolled students and visiting alumni. According to the authors, "Princeton reunions boast the second highest level of alcohol consumption for any event in the country after the Indianapolis 500. The social norms for drinking at the university are clear: Students must be comfortable with alcohol use to partake of Princeton social life."

First, the authors had students complete questionnaires about their private feelings around drinking at Princeton, questions like, "How comfortable do you feel with the alcohol drinking habits of students at Princeton?" and "How comfortable does the average Princeton undergraduate feel with the alcohol drinking habits of students at Princeton?" In the second iteration of the study, they tweaked their questionnaire and added an extra question that asked students to rate their friends' attitudes about drinking. The authors found that individual students expressed discomfort with drinking on campus, but they also believed the average Princeton student, as well as their friends, was more comfortable with drinking on campus than they were.

Next, they asked second-year students the same questions, then interviewed them again eight weeks later. The results bore out their prediction that people would internalize their misperceptions of the community norms and bring their own behavior into line. Turns out men and women experienced the same levels of ignorance around substance use, but their responses were very different. The men changed their behavior to adhere to what they believed the community norms about drinking to be, but women did not.

Women, when faced with the idea that others value drinking more than they do (even though this belief is untrue), feel alienated and disengage from the group, and that alienation increases over time.

Finally, the authors capitalized on their timing: Princeton's president had recently instituted a campus-wide ban on kegs of beer as a way to pay lip service to the many people who were concerned about the high rate of alcohol use on campus. As soon as it was announced, the ban was met with massive disapproval, but the authors suspected that students' private attitudes might not be as negative as the public outcry. The authors' survey found that overall, students were convinced everyone else on campus had stronger negative emotions about the keg ban than they did.

The consequences of these findings are significant and worth discussing before kids head off to college. Perception shapes reality: alcohol continues to play a key role in college social life not because students want or need alcohol to be a part of the festivities, but because the individual students organizing the festivities believe other students will want to drink and expect it to be available. This also means that alcohol-free options will remain rare because kids (or staff) mistakenly believe that no one will come if alcohol isn't being served. Finally, people who feel alienated are less likely to act to change group norms (think about the kid who does not understand the lesson but won't raise his hand).

Adolescents hate being manipulated against their will (I know, duh) and they absolutely do not want to feel alienated (who does?), so they are predisposed to hear me when I explain how misunderstanding societal beliefs can make them change their behavior (men) and cause them to feel alienated from their peers (women). It definitely helps to empower them with knowledge about the concept of pluralistic ignorance and arm them with objective facts about substance use on campus.

College Drinking, Defined

Before we get into the topic of how often, and how much, college students drink, we must define our terms. You might want to sit down for this, because our perception of "a drink" and the industry and medical standard for "a drink" can be two very different things. The standard size for "a drink" depends on the type of alcohol. For beer, it's twelve ounces (a standard can or bottle, and for the record, the classic Solo cup holds sixteen ounces); for wine, it's five ounces (a standard restaurant pour is five, but at home, people tend to pour at least seven to nine ounces); and for hard alcohol, it's 1.5 ounces. These measures are not random; they reflect the amount of alcohol an average human body can process in one hour. Another myth you can bust: the lines inside Solo cups are not "smart drinking lines," as some internet sites have claimed. Snopes.com busted this one a long time ago: quoting a spokesperson for Dart Container Corporation, "The lines on our Party Cups are designed for functional performance and are not measurement lines. If the lines do coincide with certain measurements, it is purely coincidental."

When we talk about college drinking habits, we have to talk about binge drinking, defined as five drinks in a row in one episode of drinking for men, and four for women. While the number of college students who drink has hovered around 44 percent for years, and the statistical average number of drinks per week for college students is five, these numbers are deceptive because the total amount of alcohol drunk on college campuses is not evenly distributed among that 44 percent of students, and they are not drinking just five drinks per week. The heaviest-drinking college students consume 68 percent of all the alcohol drunk on campus, and the heaviest binge drinkers chug 72 percent of all alcohol consumed by college students. Binge drinking is, as my Butterfield resident discovered, incredibly dangerous for the drinker and negatively impacts the larger college community. According to Henry Wechsler, principal investigator of Harvard School of Public Health's College Alcohol Study, "Compared to non-binge drinkers, frequent bingers are 17 times more likely to miss a class, ten times more likely to vandalize property, and eight times more likely to get hurt or injured as a result of their drinking. Binge drinking also contributes to poor academic performance and college dropout rates." Women tend to underestimate their drinking problems more often than men do, and when they binge, women are at particular risk of physical harm: "About 10 percent of female students who are frequent binge drinkers report being raped or subjected to nonconsensual sex, compared to only 3 percent of non-bingeing female students," writes Wechsler.

Most Likely to Kill the Keg

Who are these binge drinkers? White men under twenty-four. Students who live in the Greek system (four out of five students who live in fraternities and sororities are classified as binge drinkers). Sorority and fraternity presidents (Greek leaders not only drink the most, they also set the norms for drinking within the house). Student athletes (29 percent of student athletes are binge drinkers, but again, this figure is deceptive because rates are disproportionately higher among athletes in high-contact sports: football, hockey, lacrosse, and wrestling). Kids who intend to drink in college, who rank parties as important, and who look forward to and romanticize drinking culture, drink more in college. Finally, students whose personalities can be described as pleasure-seeking, extroverted, impulsive rebellious, and nonconforming drink more often, in greater amounts, and with more negative consequences than other college students.

As for the rest of the kids on campus? They don't drink very much. Fifty-six percent of students on campus do not binge drink, and consume just 9 percent of the alcohol consumed on campus. The students least likely to binge are African Americans (especially African American women), Asians, students over twenty-four, students who are married, students who prioritize academics or activities such as music or art, and students who volunteer their time. Students at religion-oriented colleges and universities and armed-forces-affiliated institutions have much lower levels of alcohol consumption. The colleges and universities with the lowest rates of alcohol consumption bear these associations out: Brigham Young University, College of the Ozarks, Thomas Aquinas College, United States Naval Academy, Wheaton College, Calvin University, Gordon College, United States Air Force Academy, Grove City College, and City University of New York–Baruch College.

The Happy Drunk and the Sad Drunk

Now that we know the who, let's talk about the why, because knowing why kids drink can help us target our preventions and interventions more effectively. For example, we know that when kids drink to cope emotionally, rather than for social reasons, they are more likely to drink more heavily and face more negative consequences as a result of all that self-medication. In surveys, college students report they drink to relieve stress, dull inhibitions in order to become more extroverted or socially competent, to cope emotionally, for sensation-seeking, and because they are impulsive and don't think before they drink. That said, personality matters, and some personality types drink more in college than others. This is important information for parents because if we know that our kids may be at higher risk, we can adjust our pre-college discussions to include that information. The more kids know about themselves and how that may affect their propensity to drink, the more objective they can be about their decision making and the more they can use their knowledge to formulate counterarguments to peer pressure.

Now that we are all inordinately freaked out about binge drinking, let me drop some good news as a palate cleanser. If your child does binge drink in college, there's a good chance they will leave it behind after graduation (save for those Princeton reunions, I suppose). There's a famous study of soldiers in the Vietnam War who used heroin while they were on duty because the drug was readily available, cheap, and socially acceptable as a way to cope with the pain, discomfort, trauma, and boredom of war. Almost half of the soldiers studied admitted to having used narcotics while in Vietnam and 23 percent had positive drug tests upon departure from the country. After they returned to the United States, however, nearly all of the soldiers stopped using narcotics, even though 10 percent of them tried heroin again at home. Their dependence was situational, in response to the stress of wartime, and most were able to walk away from the drug without help. Likewise, the research on drinking patterns among heavy college drinkers mirrors this phenomenon: binge drinking among members of fraternities and sororities, the heaviest drinkers on campus, does not tend to persist after graduation. Research also indicates that kids who drink for social reasons (as opposed to emotional reasons), even those who drink heavily, are less likely to continue to drink heavily or experience consequences of drinking after college. Of course, I'm not saying parents should not worry about their heavy-drinking children, and if kids drink for emotional reasons it would be wise to intervene sooner rather than later. According to the research, stress- or anxiety-based drinking is associated with long-term drinking and increased rates of negative outcomes.

And here's more good news: parents continue to matter, even in college. Our words and, more important, our actions and attitudes continue to influence our young adults even after we wave good-bye on college drop-off day. Here's how parents can best use the time before college starts to prepare kids for their transition from high school to college, and college to the world.

How to Talk Using the Right Tone

Conversations about substance use should be part of a lifelong discussion around healthy behaviors and habits, but even if you are getting started late, here are a few ways to grease the conversational wheels:

- BE PREPARED WITH DATA. Parents are a young adult's primary source of information about health and healthy behaviors, so we'd better know what we are talking about. The most recent surveys show that alcohol and drug use are declining across adolescence and young adulthood, but most college students believe their peers are using more than they actually are, and further, they believe their peers care more about using than they actually do. Because perception drives reality, it's important that parents help their kids bring perception in line with reality. Have some statistics at the ready in case your child is misperceiving her peers' attitudes and habits. You can find the most up-to-date statistics on adolescent attitudes and behaviors regarding drug use in Monitoring the Future, a report published annually by the National Institute on Drug Abuse.

- DON'T LECTURE. Lectures don't work in school-based substance abuse prevention programs and they definitely don't work when discussing alcohol and drugs around the dinner table. Kids tell me all the time that they don't really have a problem talking about alcohol and drug use, but they definitely don't want to be lectured to.

- ASSUME A TONE OF RESPECT AND TRUST. When adolescents feel they are respected and trusted to make good decisions, they are more likely to warrant your respect and make good decisions. Balance admonitions and warnings with praise; if they make a good decision, tell them how proud you are.

- BE AWARE OF DISCUSSION DEAD ENDS. I've been informed by one of my sons that I have a tone he really resents, one he describes as a "Don't even bother, I know more than you" tone that makes him feel so disrespected he won't engage once he's heard it. It took me a while to understand what he was hearing, and even longer to stop using it, but now that it's been excised, our discussions about touchy topics go much more smoothly. Other conversational dead ends include ad hominem attacks or value judgments such as "Well, (he, she, it, that) was (stupid, wrong, evil)."

- LISTEN, EVEN WHEN YOU DISAGREE. Not everyone speaks in fully formed, meticulously thought-out statements. Allow for some silent pauses, and then reflect back what you heard. There's always the possibility that's not what was meant and often, hearing a flawed rationale repeated back to them can be just the thing to help reframe and adjust their perspective. Sometimes the statement, "I want to rush a frat next year," is not a decision, but an opportunity to put a thought out into the ether and check out how it sounds. Instead of offering your opinion on the thought, your response might be, "Huh, that's interesting. Why a frat?"

- RAMP UP THE TALK IN THE SUMMER BEFORE FRESHMAN YEAR. According to multiple studies of both average and high-risk students, it's important to talk about substance use in the months before school starts, then offer booster shots of these discussions throughout college.

- BE FAMILIAR WITH YOUR CHILD'S HOPES, DREAMS, GOALS, AND PRIORITIES. If you understand these things, you not only show your kid you have been listening and learning about them, you possess powerful behavioral levers and opportunities to open conversations about how drugs and alcohol can get in the way of achieving what they want from college and life.

- DON'T BE AFRAID TO DISAGREE. It happens, and it can even be a natural starting place for eventual compromise. Try not to freak out when your kid does not swallow your ideas, ethics, or values whole cloth, because freaking out is the fastest way to solidify their tentative disagreement into flat-out opposition. Listen, reflect back what you hear, and have them explain from their perspective.

What to Talk About

Now that we know how to talk, here are some ways into a conversation about substance use:

- WHAT COLLEGE WILL THEY ATTEND? Every college has its individual culture. If a school appears on *Princeton Review*'s list of "Party Schools" (Syracuse University, University of Alabama at Tuscaloosa, University of Delaware, West Virginia University, Tulane University, University of California, Santa Barbara, Colgate University, Wake Forest University, Bucknell University, University of Rhode Island) or "Lots of Hard Liquor" (University of California, Santa Barbara, Syracuse University, Wake Forest University, Colgate University, University of Delaware, Tulane University, University of Maine, Bentley University, Grinnell College, Elon University), "Lots of Beer" (University of Wisconsin–Madison, Union College, Eckerd College, West Virginia University, Colgate University, Wake Forest University, Tulane University, Syracuse University, University of Alabama at Tuscaloosa, University of Delaware), or "Reefer Madness" (University of Vermont, University of Rhode Island, Skidmore College, Reed College, University of Maine, Bard College, Marlboro College, University of California, Santa Barbara) one can assume drug and alcohol usage rates will be higher than average. If my kid was planning to apply to or attend any of these schools (and, full disclosure, one of my sons might just) I will talk with him about whether, given all other factors, he wants to attend a school known for its high drug and alcohol use. School rules and reputation drive consumption, but so does location. One study, "The State Sets the Rate," found that the higher the rate of binge drinking in a state, the higher the rates of binge drinking among college students tend to be.

- **WHERE THEY PLAN TO LIVE DURING THE SCHOOL YEAR AND WHY.**
  Where kids live during college is associated with the amount they drink, either because students self-select for environments where alcohol is prioritized or because the culture in that en-

vironment fosters use. Students who drink the most live (in descending order, from most alcohol consumed to least) in fraternities and sororities, in off-campus housing, in standard campus housing, in designated healthy-living or substance-free housing, and at home with family. Students affiliated with the Greek system drink more often, drink more each session, and incur more negative consequences from drinking. In fact, the intention to live in a fraternity or sorority when they eventually go to college has been associated with higher drinking levels among high school students. If the frat dream pops out of your kid's mouth, explore their reasons. If your kid has lived in a dorm and suddenly opts to move off campus, talk about that, too.

• WILL YOUR CHILD BE A STUDENT ATHLETE OR HARD-CORE SPORTS FAN? College athletes drink more heavily, more frequently, and experience more negative consequences than nonathletes. However, some sports engender more substance use than others. Generally speaking, the higher the physical contact in the sport, the higher the rates of drinking. Football, lacrosse, hockey, and wrestling top the substance use charts, while individual, no-contact sports tend to have lower levels of substance use. Among heavy-using sports teams, captains and team leaders, especially males, tend to drink more than their teammates (just as leaders tend to be the highest drinkers in fraternities and sororities). Turns out, prioritizing sports in college, even among nonparticipants, is a risk factor for heavier use. Students who respond positively to the survey question, "athletics are important" tend to have higher rates of alcohol use.

• DOES THE COLLEGE OFFER SPECIAL HOUSING OPTIONS? This can be as easy as asking, "Did you look on the college's residential life page to see what other housing options there are other than standard freshman dorms?" Frankly, a lot of freshman dorms stink, literally and figuratively (freshmen are often given the least-requested housing), but increasingly, colleges and universities offer special housing including substance-free halls and dorms (sometimes marketed more broadly as healthy living residences or wellness environments). Many offer exercise classes, fitness centers, healthier food options, single rooms, expanded quiet hours, and incentives for healthy living such as activity-tracking devices in addition to promoting substance-free living. Most require residents to sign an agreement to not possess alcohol, tobacco products, or drugs in the residence hall. Some colleges and universities even offer sober recovery housing for students committed to their sobriety.

- THEY WILL ASK ABOUT YOUR BEHAVIOR AND USE IN COLLEGE. Best to be prepared on this front. Don't be afraid to write your answers down to think about them. If you used drugs and alcohol, don't glamorize or romanticize your use. You are not pitching the plot for a bromantic comedy like *The Hangover*, you are showing your kid that you can empathize with the temptation to use while supporting smart decision making. In our home, we talk about why I almost always chose sobriety in college and how that helped me get what I wanted out of my education. My husband made different decisions, some he regrets, and we talk about those, too. What opportunities were lost, what he wishes he'd done differently. Your experience—your highs, lows, and mistakes—can be useful information, as long as you are not handing them an instruction manual for debauchery.

- REVISIT YOUR SCRIPTS FOR DECLINING DRUGS AND ALCOHOL. One of the most effective ways to prepare kids to say "No, thanks,

I'm good" is to practice saying "No, thanks, I'm good." Knowing you can say it and being able to say it when it feels like everyone else in the room is saying yes are two very different things. This technique, one that's proved to work in multiple studies, is called practicing refusal, rehearsing refusal, and inoculation messaging, but whatever you call it, it can proactively protect kids from feeling the pressure to use. There is a list of scripted lines and excuses in chapter 8.

- DO SOME MORE MYTH BUSTING. No, coffee does not sober you up, nor does a cold shower, walking, slapping someone in the face, drinking milk, placing a penny in your mouth, eating, or exercise. You can't render someone less drunk or more able to drive using any of these methods. The body is only capable of processing alcohol at the rate of approximately one drink per hour, and there's no magic formula to speed that up. There are, unfortunately, many ways to slow it down or compound alcohol's effect in the body. Many common drugs lose their potency when mixed with alcohol, have dangerous side effects that are compounded when mixed with alcohol, or have the potential to interfere with the liver's ability to process either the medication or the alcohol or both. These include antibiotics, acetaminophen, aspirin, ibuprofen, high blood pressure medications, narcotics, sedatives, and tranquilizers and many others. Tell them to always take a look at the contraindications on the drug label to be sure.

- DRINKING IS ILLEGAL FOR STUDENTS UNDER TWENTY-ONE. It seems insane to have to mention this, but colleges continue to turn a blind eye to underage drinking. Administrations in colleges and universities tend to view underage drinking as an expected and regular violation, and don't bother to enforce consequences. If you expect your child to be supervised or held accountable for their underage drinking, I'd adjust those expectations. I'd also refrain from telling your child that there will be consequences for their underage drinking in college, because it's not true and saying so undermines your credibility while highlighting just how out of touch you are with reality. I asked my underage son how many times he'd have to get caught with an open container at his college to face any sort of consequences, and he replied, on various occasions, "five" and "seven." I did a deep dive into the college's alcohol policy and while I could not find evidence of either number (the rules on "warnings" and "sanctions" seem pretty open to individual interpretation) I think it's notable that *he* believes the answer is five or seven. As I've mentioned a couple of times, perception drives reality where drug and alcohol use is concerned.

- REVISIT DISCUSSIONS ABOUT THE WAYS ALCOHOL AND DRUGS AFFECT THE BODY. Review the information in chapter 3 about what drugs and alcohol do to the brain and why they are more harmful in people under the age of twenty-five than in adults whose brains are fully developed. Any time you can interject facts into your discussions of drugs and alcohol, you are not only arming your child with reasons and excuses not to drink or use, you also reinforce their baseline attitudes about the role of drugs and alcohol in their life.

- REMIND THEM OF GENDER DIFFERENCES. Women are at increased risk of alcohol poisoning due to their body composition, the way they process alcohol, and their smaller size. Drinking games (I will get there in a minute) are especially risky for women in that there's no handicapping for gender and keeping up with the boys can be a very dangerous strategy.

- TALK ABOUT THE SPECIAL RISKS OF DRINKING GAMES. Most colleges and universities prohibit drinking games and many even
  go to the effort to provide examples in their handbook for the
  purposes of clarification, but again, these policies are not often
  enforced. Whether we are talking about keg stands, Beer Pong,
  Quarters, timed interval drinking (PowerHour, or drinking
  every minute on the hour is a popular version), Beer Olympics, Flip Cup, Drunk Jenga, Drinkapalooza, Beer Roulette,
  Edward 40 Hands, or many, many others, they all promote
  binge drinking, drinking fast, and drinking to get drunk and
  are extremely dangerous. If you are curious which games your
  kid will be most likely to encounter at a given college or in
  a given state, CollgeStats.org published an exhaustive survey
  on the topic, collected via Instagram mentions, at "Drinking
  U." For the record, Beer Pong is the most popular game at the
  Massachusetts Institute of Technology and the University of
  Scranton while Flip Cup rules at Rutgers and Yale.

I spend a fair amount of time on college campuses because I
am invited to speak to visiting parents about my first book, *The
Gift of Failure*, as part of the scheduled festivities. I love hanging
out on campus before my event, watching rosy-cheeked students
show their parents around, pointing out where they study, eat, and
attend classes. Recently, I visited my own son during his college's
family weekend, with my younger son, who was fifteen at the time,
and my husband. My son lives in a college-owned house with
about twenty other kids, and they invited visiting family to a party
at the house. They purchased food and drinks, and grilled burgers
and hot dogs for dinner. We arrived late, and when we entered the
living room we were greeted by the sight of a line of parents standing along the Beer Pong table. I could call the table a dinner table

or table tennis table, but in all honesty, it's the Beer Pong table. As I moved into the room, I realized they were doing ski shots. The game is ubiquitous in New England and fairly simple: five people drink in unison from shot glasses affixed along a ski. I was a ski instructor in New Hampshire for a few years after college and every rented condo or house I have ever been in had a ski-shot ski at the ready or on prominent display, usually over the fireplace mantel.

I'm not going to lie: it was a weird and touching evening. My son's housemates were anxious to be good hosts for their parents while bonding for the first time over shared college traditions and experiences. Meanwhile, the parents were anxious, too. The pressure was on to not appear tragically uncool and embarrass their children in front of their friends and the other parents. As I have absolutely no problem being tragically uncool, I sat back and watched the evening unfold.

The Beer Pong (and ski shot) table was at the literal and metaphorical center of the evening, and at first, I saw this as a clear example of how permissive parenting attitudes around alcohol contribute to adolescent drinking. If, as research shows, college drinking is driven by parent and community attitudes, perceptions, and norms, and further, that leaders (in this case, parents) have the most power to shape those attitudes, perceptions, and norms, then yes, this entire evening freaked me out.

However, as I made polite conversation with other parents and watched the kids run around, trying to encourage their parents to do ski shots and play Beer Pong, I realized just how complicated the interplay of influences and pressures in the room really was.

Many of the parents declined to participate, and they offered their excuses freely and unabashedly. They explained they don't drink, had to drive to the hotel, or had promised to get up early in the morning for the Spin class their kid insisted they attend. The parents who did participate, however, were clearly doing it to make their kids proud. Many opted to fill their shot glasses with wine or beer instead of whiskey, and one mom even filled hers with water, but gosh darn it, they were not going to let their kids down.

I'm still processing the events of that evening, untangling the various forces at work and my place within them. I was not drinking, of course, but I was there. I knew my son and many of the students in that room were not twenty-one. I allowed my fifteen-year-old to be there, watching. Later, my husband admitted he'd found the dynamics challenging as well, and recounted a conversation he'd had with another parent. They were discussing the presence of younger siblings at family weekend, and the parent said, "When I dropped my son off to visit his sister at college, we wanted him to have fun, to see what it was like. I said to her, 'I'm not going to freak out if I hear he gets drunk. That's not the issue. Keep an eye on him. Keep him safe. Make sure he doesn't do anything that hits the evening news.'"

Except, getting drunk is precisely the issue, and the way we feel about it, talk about it, and manage it in our parenting informs the way our children will feel about it, talk about it, and manage it as they move out into their own lives.

In my favorite photograph from that night, three kids and two parents are doing a ski shot. The dad in the foreground, drinking from the tip of the ski, is wearing a shirt emblazoned with the word "Thirsty?" My kids are in the background, watching. My older son is focused on the drinkers and laughing with two of his housemates, but my younger son is looking at me, directly into the camera, as I take the picture.

That kid may be focused on me now, but he will head off to college in less than two years. In the meantime, I plan to make the most of the precious time I have left in his waning, ever-wandering gaze.

CHANGING THE ENDING

O ver my twenty years as a teacher and a parent, I've watched a lot of students graduate. They head off into whatever comes next: high school, college, or adulthood. Once a year, I dressed up, cried a little, shook a lot of hands, and said congratulations to my students and their parents.

Most of the kids I've taught take graduation for granted. I know I did. Many came from privileged backgrounds, complete with the resources, support, health care, and financial backing to go anywhere and do anything. They were expected to graduate, to go off into the world via paths worn smooth by their parents' footsteps. They matriculated to elite schools and later, to elite careers.

I've always been proud of my students; they've gone on to become teachers, surgeons, artists, professors, mathematicians, and translators. I'm honored to have been a part of their education and am happy to share in the reflected glow of their success. Recently, however, I've come to value the quieter, harder victories even more than big wins heralded by pomp and circumstance.

My favorite graduation ceremony of all was held last year, a few months before the rehab closed to adolescent clients. I did not yet know for sure that I would be losing my teaching job, but everyone on the faculty suspected changes were afoot and the end was near.

I arrived at the rehab one Friday morning to find that my class had been canceled in favor of a hastily arranged graduation ceremony for Jeremy, a seventeen-year-old boy whose parents had just arrived to pick him up.

Rehab graduation is an informal yet solemn ceremony. The kids, their counselors, the unit director, and parents or guardians sit in a circle, and pass around a graduation coin as each person shares their hopes, fears, and a favorite memory of their time with the graduate.

Hopes come first, usually offered as advice for continued sobriety, happiness, and success.

"Stay out of trouble."

"Be good on probation."

"Take this shit seriously."

"Graduate from high school."

"Remember how good it feels to come this far."

"Don't give in to the 'Fuck-its.'"

"Have some emotional regulation, man."

"Don't put metal in any outlets."

"Be a leader, not a follower."

Fears are trickier sentiments, often because they are rooted in graduates' flaws, the character traits or behavioral patterns that landed them in rehab in the first place.

"Don't get pulled in by fake friends who don't have your best interest at heart."

"Try not to lose control, because when the gloves go off, you're gone, and your ability to listen to reason goes away."

"Man, think before you go back to stealing cars and doing other stupid shit. That's not you. You are better than that."

"Stop before you do that first dumb thing."

"Remember the world isn't against you."

"Don't shut people out. Talk to the people who love you even when you fuck up."

"I'm so scared you will listen to that little voice in your head that says you can use just a little bit and it will be okay because it won't. It won't be okay."

It's the final offering, the memories, that brings the house down. In these last moments together, a sense of urgency prevails. Defenses collapse and hearts open.

"I will always remember that every morning at wake-up, you wiggled your foot to let me know you were awake and heard me. Made me smile every time."

"That first night, when we poured cold water on you and threw tissues at you and laughed so hard. That's when I knew you were cool."

"I was so proud of you when you applied to all those jobs, and that day you got a call back? I will always remember the look on your face. You were so proud of yourself."

"You were there for me."

"That time you gave me tea. It was good tea and I liked it. Thank you for that."

"I will remember making you laugh. I love making people laugh, but especially you."

"You offered to clean my shoes. Man, that was so nice."

Finally, the coin makes its way into Jeremy's hands. He sat between his mother and father, a kind and nurturing couple who adopted him after struggling to have their own child. He looked down at the coin in the palm of his left hand, a token reminder of everything he'd learned and the progress he stood to lose.

Turning to his father, he said, "When I was three or four years old, I wanted to be a policeman just like you, so you made me a uniform and played police with me. I'm so glad you are my dad."

And to his mother, he said, "When I was three or four I used to think there were monsters in my room, and you would wake up and come in and hide under the covers with me when I was scared. I would not have been able to cope with the monsters without you. You are the best mother I could have had."

Jeremy passed the coin to his father, looking expectant and hopeful. His father took a deep breath and held it for a little longer than we all expected. The room grew tense.

"I'm so afraid for you," he admitted. "I fear your friends, your impulsivity, and your anger. There are a lot of things I wish I'd done differently, and I know it's going to be a long road, but I am so proud of you."

Jeremy leaned in for a hug but his father did not offer one. He patted Jeremy on the upper arm and passed the coin to his wife.

His mother waved the group off at first, as if to forfeit her turn to speak. After a moment to collect herself, though, she said, "You are my little boy. We've been through hell. We hated each other. I look at you and I love you more than I can ever say. You are my son. I am so proud of you. I want to see you sober. You are the son I always wanted. I fear that I will lose him. Don't let me lose him."

As I hugged Jeremy and wished him well, I yearned to offer reassurances, a solemn promise that a sober today would lead to a sober tomorrow, that all of our hopes for him would come to pass and all of our fears would prove unfounded. Because I could not guarantee any of that, I gave him my copy of his favorite graphic novel. It was the best I could do.

That's all we can ever do for the children we love.

A few weeks after the rehab graduation ceremony, I discovered by way of Facebook that Georgia, now thirty-six, finally graduated from college. She posted a picture to her wall: post-ceremony, wearing her black cap and gown, surrounded by her mother, father, and her girlfriend, Shauna. Her parents look relieved and cautiously optimistic. Her girlfriend, Shauna, beams with pride.

When Georgia left my classroom for the last time, I'd given her a book, too. She'd loved our readings in Walt Whitman's *Leaves of Grass*, so I'd taken to calling her "My Little Spider," as described in the poem "A Noiseless Patient Spider":

A noiseless patient spider,
I mark'd where on a little promontory it stood isolated,
Mark'd how to explore the vacant vast surrounding,

It launch'd forth filament, filament, filament, out of itself,
Ever unreeling them, ever tirelessly speeding them.

And you O my soul where you stand,
Surrounded, detached, in measureless oceans of space,
Ceaselessly musing, venturing, throwing, seeking the
    spheres to connect them,
Till the bridge you will need be form'd, till the ductile
    anchor hold,
Till the gossamer thread you fling catch somewhere, O
    my soul.

That was Georgia at seventeen: casting threads about in all directions, desperately hoping to latch on to something—anything—
that would hold her fast to this world and rescue her from the
isolation on the lonely promontory she'd made for herself.

I kept reaching out, grasping at every thread, believing I was
supposed to save her from herself and keep her from the big bad
world of pain, and loss, and hurt. In the end, I could not, nor could
anyone else. All we could do was love her and promise we'd return
to her side, relieved and cautiously optimistic, when she was ready
to save herself.

I printed that Facebook photo out so I could pin it to the wall
near my desk, the place I go when I need an emotional or motivational boost. Staring out from my wall of inspiration, Georgia
looks healthier and more beautiful than I have ever seen her.

Most clearly, though, Georgia is happy.

ACKNOWLEDGMENTS

When I stumbled into teaching more than twenty years ago I had no idea it would bring me so much joy, hope, and love. I certainly could not have known way back then, well before my alcoholism had taken hold, that teaching in my small rehab classroom in the woods of Vermont would keep me sober.

To the students of the Duke Talent Identification Program, Rowland Hall/St. Mark's School, Highland High School, Hanover High School, Crossroads Academy, and Valley Vista, thank you for all you have given me.

You have, quite literally, saved my life.

NOTES

Chapter 1: Hi, My Name Is Jess, and I'm an Alcoholic

5   **progression from alcohol use to dependence** J. Smith and C. Randall, "Anxiety and Alcohol Use Disorders: Comorbidity and Treatment Considerations," *Alcohol Research: Current Reviews* 34, no. 4 (2012): 414–31.

17   **"our family's well-worn slippery slope"** "Coming Out to My Children About My Alcoholism," *HuffPost*, August 24, 2013, https://www.huffpost.com/entry/coming-out-to-your-children-about-alcoholism_b_3480217.

Chapter 2: A Long, Strange Trip: Drugs, Alcohol, and Us

26   **"very few try illicit drugs other than marijuana"** D. Kandel and K. Yamaguchi, "From Beer to Crack: Developmental Patterns of Drug Involvement," *American Journal of Public Health* 83, no. 6 (1993): 851–55.

26   **"greatly increases the likelihood"** S. Nkanasah-Amankra and M. Minelli, "'Gateway Hypothesis' and Early Drug Use: Additional Findings from Tracking a Population-Based Sample of Adolescents to Adulthood," *Preventative Medicine Reports* 4 (2016): 134–41.

27   **White males are more likely to escalate their use to hard drugs** H. White et al., "Stages and Sequences of Initiation and Regular

Substance Use in a Longitudinal Cohort of Black and White Male Adolescents," *Journal of Studies on Alcohol and Drugs* 68, no. 2 (2007): 173–81.

27 **three times as likely** Carolyn E. Sartor, "Alcohol or Marijuana First? Correlates and Associations with Frequency of Use at Age 17 Among Black and White Girls," *Journal of Studies on Alcohol and Drugs* 80, no. 1 (2019): 120–28.

28 **"If we can understand the beginning, we can help change the ending"** *The First Day: A Focus on The Beginning*, directed by Jonathan Hock (2020), https://thefirstdayfilm.com/.

28 **allowed us to organize and thrive as a species** Jeffrey P. Kahn, "How Beer Gave Us Civilization," *New York Times*, March 15, 2013, https://www.nytimes.com/2013/03/17/opinion/sunday/how-beer-gave-us-civilization.html.

30 **"the painfully shy, their angst suddenly quelled"** Ibid.

30 **"Literally, the two have gone hand in hand"** Gregg Smith, *Beer in America: The Early Years—1587–1840* (Boulder, CO: Siris Books, 1998), 2.

31 **left their homes for the uncertain promise of a new land** Mark Edward Lender and James Kirby Martin, *Drinking in America: A History* (New York: Free Press, 1987), 5.

31 **"pumpkins, and parsnips, and walnut-tree chips"** Eric Burns, *The Spirits of America: A Social History of Alcohol* (Philadelphia: Temple University Press, 2004), 13.

31 **pre-Pasteur and his germ theory** Pasteur was not born until 1822 and did not conduct the experiments that led to his germ theory until 1860–64.

32 **"will drink one after another until they have emptied them"** Christopher M. Finan, *Drunks: An American History* (Boston: Beacon Press, 2017), 11.

32 **the highest of any population group in the United States** Substance Abuse and Mental Health Services Administration report on American Indian /Alaska Native data, accessed March 2, 2020, https://www.samhsa.gov/sites/default/files/topics/tribal_affairs/ai-an-data-handout.pdf.

33 **tavern adjacent to the Lexington Green** Smith, *Beer in America*, 98–99.

33 **"I have frequently seen Fathers wake their Child"** W. J. Rorabaugh,

The Alcoholic Republic: An American Tradition (Oxford: Oxford University Press, 1979), 14.

33 **"It is no uncommon thing"** Ibid.

33 **"Everyone drank, beginning at birth"** Susan Cheever, *Drinking in America: Our Secret History* (New York: Hachette, 2015), 34.

34 **"toddies for toddlers"** Ibid.

34 **"lethal lullabies"** M. Obladen, "Lethal Lullabies: A History of Opium Use in Infants," *Journal of Human Lactation* 32, no. 1 (2016): 72–85.

35 **remained on the shelves** American Medical Association, "Nostrums and Quackery: Articles on the Nostrum Evil and Quackery Reprinted from the Journal of the American Medical Association, Volume 2" (Chicago: Press of American Medical Association, 1921; Google Books).

35 **"If it were not for [opium] and my soda-water"** John C. Burnham, *Bad Habits: Drinking, Smoking, Taking Drugs, Gambling, Sexual Misbehavior, and Swearing in American History* (New York: New York University Press, 1993), 115–16.

35 **"Indeed, fears about young people and drug use"** Alex Mold, "The Historical Context of Drug Use by Young People," in *Substance Misuse and Young People: Critical Issues*, ed. Ilana B. Crome and Richard Williams (New York: Routledge, 2020), 9.

35 **"Meth[amphetamine] was considered"** Ryan Grim, *This Is Your Country on Drugs: The Secret History of Getting High in America* (Hoboken, NJ: Wiley, 2009), 52.

36 **"renewing their interest in life and living"** Jerrold Winter, *Our Love Affair with Drugs: The History, the Science, the Politics* (New York: Oxford University Press, 2020), 66.

36 **80 percent of them to women** Grim, *This Is Your Country on Drugs*, 52.

36 **"minimal brain damage"** Nicolas Rasmussen, *On Speed: The Many Lives of Amphetamine* (New York: New York University Press, 2008), 164.

36 **48.18 *metric tons*** Brian J. Piper et al., "Trends in Use of Prescription Stimulants in the United States and Territories, 2006 to 2016," *PLoS ONE* 13, no. 11, https://journals.plos.org/plosone/article?id=10.1371/journal.pone.0206100.

37 **"all-out offensive" against both the supply and demand** "President Nixon Declares Drug Abuse 'Public Enemy Number One,'" presidential address given on June 17, 1971, YouTube video posted

April 29, 2016, https://www.youtube.com/watch?v=y8TGLLQlD9M, accessed March 1, 2020.

39   **speakeasies in New York City** Burns, *The Spirits of America*, 197.

41   **"Pretending that things are not as they seem"** Susan Cheever, *Note Found in a Bottle: My Life as a Drinker* (New York: Washington Square Press, 1999), 187.

41   **"the word 'dependence' won over 'addiction' by a single vote"** Charles O'Brien, "Addiction and Dependence in DSM-V," *Addiction* 106, no. 5 (October 2013), https://www.ncbi.nlm.nih.gov/pmc/articles/PMC3812919/.

Chapter 3: Wired for Risk: A Primer on the Adolescent Brain

46   **human mothers would have to gestate babies for 18 to 21 months** Kate Wong, "Why Humans Give Birth to Helpless Babies," Observations, *Scientific American,* August 28, 2012, https://blogs.scientific american.com/observations/why-humans-give-birth-to-helpless -babies/.

47   **eye gaze, social hierarchies, peer pressure, reputation, and physical appearance** Mary Helen Immordino-Yang, Linda Darling-Hammond, and Christina Krone, "The Brain Basis for Integrated Social, Emotional, and Academic Development," Aspen Institute and National Commission on Social, Emotional, & Academic Development, 2018, https://assets.aspeninstitute .org/content/uploads/2018/09/Aspen_research_FINAL_web .pdf.

48   **"substance abuse, unintended pregnancy, and sexually transmitted diseases"** Aaron White and Ralph Hingson, "A Primer on Alcohol and Adolescent Brain Development: Implications for Prevention," in *Prevention of Substance Use,* ed. Zili Sloboda, Hanno Petras, Elizabeth Robertson, and Ralph Hingson (Switzerland: Springer Nature, 2019), 8.

49   **"Plasticity is the process through which"** Laurence Steinberg, *Age of Opportunity: Lessons from the New Science of Adolescence* (Boston: Houghton Mifflin, 2014), 24.

54   **"You can't produce enough dopamine to get out of bed"** PBS, *Nova, Addiction,* directed by Sarah Holt. Air date October 17, 2018, https://www.pbs.org/wgbh/nova/video/addiction.

59   **"Among youth who drink, the proportion who drink heavily"** Lorena Siqueira, Vincent C. Smith, and Committee on Substance Abuse, "Binge Drinking," *Pediatrics* 136, no. 3 (September 2015), https://pediatrics.aappublications.org/content/136/3/e718.

60   **the smaller her hippocampus will be** Frances E. Jensen and Amy Ellis Nutt, *The Teenage Brain: A Neuroscientist's Survival Guide to Raising Adolescents and Young Adults* (New York: Harper, 2015), 130.

61   **psychologist Edith Sullivan, described the loss as "striking"** "'Striking' Impact of Adolescent Drinking on the Brain," Medscape Medical News, November 9, 2017. http://ncanda.org/images/Med Scape%20NCANDA%20article.pdf.

61   **there's no safe amount of alcohol during pregnancy** "Alcohol Use in Pregnancy," Centers for Disease Control and Prevention, March 27, 2018, https://www.cdc.gov/ncbddd/fasd/alcohol-use .html.

61   **alcohol can be detected in breast milk** "Alcohol," Centers for Disease Control and Prevention, December 28, 2019, https://www.cdc.gov /breastfeeding/breastfeeding-special-circumstances/vaccinations -medications-drugs/alcohol.html.

61   **"Our results show that the safest level of drinking is none"** Max G. Griswold et al., "Alcohol Use and Burden for 195 Countries and Territories 1990–2016: A Systematic Analysis for the Global Burden of Disease Study 2016," *The Lancet* 392 (August 2018), https:// doi.org/10.1016/S0140-6736(18)31310-2.

62   **"It's putting a new generation at risk"** Robert Redfield, "Progress Erased: Youth Tobacco Use Increased During 2017–2018," Press Release, Centers for Disease Control and Prevention, February 11, 2019, https://www.cdc.gov/media/releases/2019/p0211-youth-tobacco -use-increased.html.

63   **bronchitis and asthma as well as acute lung damage** Dharma Bhatta and Stanton Glantz, "Association of E-cigarette Use with Respiratory Disease among Adults: A Longitudinal Analysis," *American Journal of Preventive Medicine* 58. no. 2 (2020): 182–90.

63   **thus resulting in a net increase of new nicotine users** Stanton Glantz and David Bareham, "E-cigarettes: Use, Effects on Smoking, and Policy Implications," *Annual Review of Public Health* 39 (2018): 215–35.

64 **persistent level of memory and problem-solving deficits** Cynthia Kuhn, Scott Swartzwelder, and Wilkie Wilson, *Buzzed: The Straight Facts About the Most Used and Abused Drugs from Alcohol to Ecstasy* (New York: Norton, 2019).

67 **may lead to brain damage and loss of cognitive function** Ibid., 258.

67 **"it releases dopamine from its storage sites"** Marc Lewis, *Memoirs of an Addicted Brain: A Neuroscientist Examines His Former Life on Drugs* (New York: Doubleday, 2012), 187.

69 **chronicles with Ambien Walrus posted to Reddit's Ambien sub forum** Jamal Stone, "Insane Stories of People on Ambien That'll Brighten your Day," *MilkXYZ*, February 1, 2016, https://milk.xyz /articles/the-ambien-walrus-will-steal-your-memories/.

69 **woke up in a jail cell to the news** Allison McCabe, "The Disturbing Side Effect of Ambien, the No 1. Prescription Sleep Aid," *Huffington Post*, February 23, 2016, https://www.huffpost.com/entry /ambien-side-effect-sleepwalking-sleep-aid_n_4589743.

Chapter 4: Not My Kid: Who Gets Addicted, and Why

77 **"substance abuse disorders are complex"** Marc Schuckit, "A Brief History of Research on the Genetics of Alcohol and Other Drug Use Disorders," *Journal of Alcohol and Drugs* Supplement 17, 2014, 59–67.

78 **These environmental factors, including trauma** Bart Rutten et al., "Longitudinal Analyses of the DNA Methylome in Deployed Military Servicemen Identify Susceptibility Loci for Post-Traumatic Stress Disorder," *Molecular Psychiatry* 23 (June 2017): 1145–56.

78 **exercise** Romain Barres, "Acute Exercise Remodels Promoter Methylation in Human Skeletal Muscle," *Cell Metabolism* 15, no. 3 (2012): 405–11.

78 **sleep deprivation** Marie Gaine and Snehajyoti Chatterjee, "Sleep Deprivation and the Epigenome," *Frontiers in Neural Circuits* 12 (2018).

78 **diet** Lionel Poirier, "The Effects of Diet, Genetics and Chemicals on Toxicity and Aberrant DNA Methylation: an Introduction," *Journal of Nutrition* 132, no. 8 (2002): 2336S–39S.

78 **mental illness** "Epigenetic Mechanisms in Schizophrenia," *Bio-*

chemica et Biophysica Acta (BBA) General Subects 1790, no. 9 (2009): 869–887, https://doi.org/10.1016/j.bbagen.2009.06.009.

78   **stress** Moshe Szyf, "The Early Life Environment and the Epigenome," *Biochimica et Biophysica Acta (BBA) General Subjects* 1790, no. 9 (2009): 878–85, https://doi.org/10.1016/j.bbagen.2009.01.009.

79   **died at higher rates** Peter Ekamper et al., "Independent and Additive Association of Prenatal Famine Exposure and Intermediary Life Conditions with Adult Mortality Between Age 18–63 Years," *Social Science and Medicine* 119 (2014): 232–39, https://doi.org/10.1016/j.socscimed.2013.10.027.

79   **"and continued to do so for life"** Carl Zimmer, "The Famine Ended 70 Years Ago, but Dutch Genes Still Bear Scars," *New York Times*, January 31, 2018, https://www.nytimes.com/2018/01/31/science/dutch-famine-genes.html.

81   **"Adverse childhood experiences"** Vincent J. Felitti, "The Origins of Addiction: Evidence from the Adverse Childhood Experiences Study," English translation of article published in Germany as "Evidenzen aus einer Studie zu belastenden Kindheitserfahrungen," *Praxis der Kinderpsychologie und Kinderpsychiatrie* 52 (2003): 547–59, https://www.nijc.org/pdfs/Subject%20Matter%20Articles/Drugs%20and%20Alc/ACE%20Study%20-%20OriginsofAddiction.pdf.

82   **"70 percent Caucasian and 70 percent college-educated"** Nadine Burke-Harris, *The Deepest Well: Healing the Long-Term Effects of Childhood Adversity* (New York: Houghton Mifflin Harcourt, 2018), 39.

82   **"Our findings are disturbing to some"** Felitti, "The Origins of Addiction."

82   **most people you know have experienced one of them** Ibid.

82   **The CDC sorts them into three categories:**

Abuse

EMOTIONAL ABUSE: A parent, stepparent, or adult living in your home swore at you, insulted you, put you down, or acted in a way that made you afraid that you might be physically hurt.
PHYSICAL ABUSE: A parent, stepparent, or adult living in your home pushed, grabbed, slapped, threw something at you, or hit you so hard that you had marks or were injured.

SEXUAL ABUSE: An adult, relative, family friend, or stranger who was at least five years older than you touched or fondled your body in a sexual way, made you touch his/her body in a sexual way, or attempted to have any type of sexual intercourse with you.

Household Challenges

MOTHER TREATED VIOLENTLY: Your mother or stepmother was pushed, grabbed, slapped, had something thrown at her, kicked, bitten, hit with a fist, hit with something hard, repeatedly hit for over at least a few minutes, or threatened or hurt by a knife or gun by your father (or stepfather) or your mother's boyfriend.
HOUSEHOLD SUBSTANCE ABUSE: A household member was a problem drinker or alcoholic or a household member used street drugs.
MENTAL ILLNESS IN THE HOME: A household member was depressed or mentally ill or a household member attempted suicide.
CRIMINAL BEHAVIOR IN THE HOUSEHOLD: A household member went to prison.
PARENTAL SEPARATION OR DIVORCE.

Neglect

EMOTIONAL NEGLECT: The absence of someone in your family who made you feel important or special, loved; a family where people looked out for each other, felt close to each other, and were a source of strength and support.
PHYSICAL NEGLECT: The absence of someone to take care of you, protect you, and take you to the doctor if you needed it, who provided enough to eat; parents too drunk or high to take care of you, having had to wear dirty clothes.

83   **"Sexual abuse accounts for"** Ann Dowsett Johnston, *Drink: The Intimate Relationship Between Women and Alcohol* (New York: HarperCollins, 2013).

83   **"increase in the likelihood of becoming"** Felitti, "The Origins of Addiction."

84   **"an initial design flaw"** Ibid.

84   **adoption as a significant factor** Candy Finnigan, "Adoption before Addiction," presented to Sovereign Health Group, YouTube,

December 23, 2011, https://www.youtube.com/watch?v=IutOp
SHOeMo.

88   **"[H]ope predicted test scores and term GPA"** Shane Lopez, *Making Hope Happen: Create the Future You Want for Yourself and Others* (New York: Atria Books, 2013), 54.

90   **"34 percent of LSD use, and 30 percent of ecstasy use"** Joeseph Palamar and Caroline Rutherford, "Summer as a Risk Factor for Drug Initiation," *Journal of General Internal Medicine* (2019), https://doi.org/10.1007/s11606-019-05176-3.

Chapter 5: Tipping the Scales of Addiction:
The Protective Factors That Outweigh Risk

93   **first use typically happens in seventh or eighth grade** Thomas Ashby Wills and Grace McNamara, "Escalated Substance Use: A Longitudinal Grouping Analysis from Early to Middle Adolescence," *Journal of Abnormal Psychology* 105, no. 2 (1996): 166–80.

96   **Half of the teens who admit to misusing prescription drugs** Parnership at Drugfree.org, *2012 Partnership Attitude Tracking Study*, 2012, https://drugfree.org/wp-content/uploads/2013/04/PATS-2012 -FULL-REPORT2.pdf.

96   **Physical activity has been associated** Kimary Brener Kulig and Tim McManus, "Sexual Activity and Substance Use among Adolescents by Category of Physical Activity Plus Team Sports Participation," *Archives of Pediatrics & Adolescent Medicine* 157, no. 9 (2003): 905–12.

97   **higher among dog owners** Hayley Cutt et al., "Understanding Dog Owners' Increased Levels of Physical Activity: Results from RESIDE," *American Journal of Public Health* 98, no. 1 (2008): 66–69, https://ajph.aphapublications.org/doi/pdfplus/10.2105/AJPH .2006.103499.

98   **"the foundation of human motivation, well-being, and accomplishments"** Albert Bandura, "Adolescent Development from an Agentic Perspective," in *Self-Efficacy Beliefs of Adolescents*, ed. Tim Urdan and Frank Pajares (Greenwich, CT: Information Age, 2006), 3.

99   **can mediate the effects of poverty and prior academic failure** Frank Pajares, "Self-Efficacy During Childhood and Adolescence: Implications for Teachers and Parents," in *Self-Efficacy Beliefs of*

Adolescents, ed. Tim Urdan and Frank Pajares (Greenwich, CT: Information Age, 2006), 362.

100   "self-confident fools" Ibid., 344.

102   **First, help her understand the difference** Martin Seligman, *The Optimistic Child: A Proven Program to Safeguard Children Against Depression and Build Lifelong Resilience* (New York: Houghton Mifflin, 2007).

104   **"Children cannot be fooled"** Erik Erikson and Robert Coles, *The Erik Erikson Reader* (New York: Norton, 2000), 125.

105   **discussing their own risk behaviors** Richard Yoast et al., "Reactions to a Concept for Physician Intervention in Adolescent Alcohol Use," *Journal of Adolescent Health* 41, no. 1 (2007): 35–41.

105   **"As predicted by the adolescents"** Ibid.

106   **"Adolescents often express relief"** Sharon J. L. Levy, Janet F. Williams, and Committee on Substance Use and Prevention, "Substance Use Screening, Brief Intervention, and Referral to Treatment," *Pediatrics* 138 (2016).

110   **"the way we avoid a life of dull, boring rigidity"** Daniel J. Siegel, *Mindsight* (New York: Bantam Books, 2010), 64.

110   **lack of control over impulsivity and emotion** Rajita Sinha, "Chronic Stress, Drug Use, and Vulnerability to Addiction," *Annals of the New York Academy of Sciences* 1141 (2008): 105–30.

111   **Mindfulness practices promote** Edo Shonin and William Van Gordon, "The Mechanisms of Mindfulness in the Treatment of Mental Illness and Addiction," *International Journal of Mental Health and Addiction* 14, no. 5 (2016): 844–49.

112   **"fully present, fully awake, and alive"** M.C. Yogi, *Spiritual Graffiti: Finding My True Path* (New York: HarperOne, 2017), 90.

116   **schools start no earlier than 8:30** American Academy of Pediatrics, "Let Them Sleep: AAP Recommends Delaying Start Times of Middle and High Schools to Combat Teen Sleep Deprivation," August 25, 2014, https://www.aap.org/en-us/about-the-aap/aap-press-room/Pages/Let-Them-Sleep-AAP-Recommends-Delaying-Start-Times-of-Middle-and-High-Schools-to-Combat-Teen-Sleep-Deprivation.aspx.

117   **sleep as a cause of acne and other skin problems** National Sleep Foundation, "Teens and Sleep," accessed September 2019, https://www.sleepfoundation.org/articles/teens-and-sleep.

117   **chronic lack of sleep has been shown** Ibid.

Chapter 6: House Rules: Parenting for Prevention

122 **highest risk of developing a substance use disorder** Atika Khurana et al., "Experimentation versus Progression in Adolescent Drug Use: A Test of an Emerging Neurobehavioral Imbalance Model," *Development and Psychopathology* 27, no. 3 (2015): 901–13.

122 **twice as likely to report an intention to use drugs in the future** Center on Addiction, "Teen Insights into Drugs, Alcohol, and Nicotine: A National Survey of Adolescent Attitudes toward Addictive Substances," June 2019, https://www.centeronaddiction .org/addiction-research/reports/teen-insights-drugs-alcohol-and -nicotine-national-survey-adolescent.

123 **"more likely to report engaging in tobacco, alcohol, and marijuana use"** Ibid.

124 **used monitoring tools to check on their teens' location** Monica Anderson, "Parents, Teens, and Digital Monitoring," Pew Research Center, January 7, 2016, https://www.pewinternet.org/2016/01/07 /parents-teens-and-digital-monitoring/.

124 **"Regardless of how closely we decide to monitor"** Lisa Damour, "You Can Track Almost Everything Your Kids Do Online. Here's Why That May Not Be a Good Idea," *Time*, February 8, 2019, https:// time.com/5523239/parenting-behavior-technology-social-media/.

125 **"strong mutual attachments that persist through adolescence"** Diana Baumrind, "The Influence of Parenting Style on Adolescent Competence and Substance Use," *Journal of Early Adolescence* 11, no. 1 (1991): 58–95.

126 **"claims of their shared social norms"** Ibid.

129 **kids who sipped in fifth grade** John E. Donovan, "Really Underage Drinkers: The Epidemiology of Children's Alcohol Use in the United States," *Prevention Science* 8, no. 3 (2007), https://www .ncbi.nlm.nih.gov/pmc/articles/PMC2222916/.

129 **among sixth graders, 62 percent of boys and 58 percent of girls** Ibid.

129 **Half of European males are binge drinkers** World Health Organization, "Fact Sheet on Alcohol Consumption, Alcohol-Attributable Harm and Alcohol Policy Reponses in European Union Member States, Norway and Switzerland," 2018, http://www.euro.who.int /__data/assets/pdf_file/0009/386577/fs-alcohol-eng.pdf.

130 **"Any parent with knowledge"** Candy Finnigan and Sean Finnigan, *When Enough Is Enough: A Comprehensive Guide to Successful Intervention* (New York: Avery, 2008), 25.

130 **"clear message against the use of alcohol"** Joanna Quigley and Committee on Substance Use and Prevention, "Alcohol Use by Youth," *Pediatrics* 144, no. 1 (2019), https://pediatrics.aappublications.org /content/144/1/e20191356.

134 **impulsivity, sensitivity to rewards, sensation-seeking, and novelty-seeking** Bernard Gallo Le Foll et al., "Genetics of Dopamine Receptors and Drug Addiction: A Comprehensive Review," *Behavioral Pharmacology* 20, no. 1 (2009): 1–17.

135 **children begin to internalize** Robert A. Zucker and Stephen B. Kincaid, "Alcohol Schema Acquisition in Preschoolers: Differences between Children of Alcoholics and Children of Nonalcoholics," *Alcoholism: Clinical and Experimental Research* 19, no. 4 (1995): 1101–17.

137 **more likely to use drugs and alcohol themselves** Joseph A. Califano, *How to Raise a Drug-Free Kid: The Straight Dope for Parents* (New York: Atria Books, 2014).

137 **one out of ten teens who use addictive substances** National Center on Addiction and Substance Abuse at Columbia University, "National Survey of American Attitudes on Substance Abuse II: Teens and Their Parents," September 1996, https://www.centeronaddiction .org/.

137 **If older siblings have warm, supportive friend groups** David C. Rowe and Bill L. Gulley, "Sibling Effects on Substance Use and Delinquency," *Criminology* 30, no. 2 (1992): 217–34.

138 **lack of self-control is another trait** Kenneth J. Sher et al., "Personality and Substance Use Disorders: A Prospective Study," *Journal of Consulting and Clinical Psychology* 68, no. 5 (2000): 818–29.

138 **tobacco use can contribute** Antonio Terracciano et al., "Five-Factor Model Personality Profiles of Drug Users," *BMC Psychiatry* 8, no. 1 (2008), https://doi.org/10.1186/1471-244X-8-22.

139 **including substance abuse and risky sex** Kimary Brener Kulig et al., "Sexual Activity and Substance Use among Adolescents by Category of Physical Activity Plus Team Sports Participation," *Archives of Pediatrics & Adolescent Medicine* 157, no. 9 (2002): 905–12.

139 **"the most prevalent of adversities"** Sharlene A. Wolchik et al., "The

New Beginnings Program for Divorcing and Separating Families: Moving from Efficacy to Effectiveness," *Family Court Review* 47, no. 3 (2009): 416–35.

139 **one-third of all children** Paul R. Amato, "Children of Divorce in the 1990s: An Update of the Amato and Keith (1991) Meta-analysis," *Journal of Family Psychology* 15, no. 3 (2001).

140 **"children with divorced parents"** Ibid.

140 **in order to cope with the stress of heightened conflict** Jeremy Arkes, "The Temporal Effects of Parental Divorce on Youth Substance Use," *Substance Use & Misuse* 48, no. 3 (2013): 290–97.

143 **The rates are much lower for boys** David Finkelhor et al., "The Lifetime Prevalence of Child Sexual Abuse and Sexual Assault Assessed in Late Adolescence," *Journal of Adolescent Health* 55, no. 3 (2014): 329–33.

143 **parental alcohol abuse increases a child's risk of experiencing sexual abuse** Cathy Widom et al., "Alcohol Abuse as a Risk Factor for and Consequence of Child Abuse," *Alcohol Research & Health* 25, no. 1 (2001): 52.

143 **girls drink slightly more once they get started** Sharaf Khan et al., "Gender Differences in Lifetime Alcohol Dependence: Results from the National Epidemiologic Survey on Alcohol and Related Conditions," *Alcholism: Clinical and Experimental Research* 37, no. 10 (2013).

144 **further to fall when high than their male counterparts** Krista Lisdahl Medina, "Imaging Study: Prefrontal Cortex Morhometry in Abstinent Adolescent Marijuana Users: Subtle Gender Effects," *Addiction Biology* 14, no. 4 (2009) 457–68.

145 **women are more likely than men** Huw Thomas, "A Community Survey of Adverse Effects of Cannabis Use," *Drug and Alcohol Dependence* 42, no. 3 (1996): 201–7.

145 **may occur more rapidly among women than men** Carlos A. Hernandez-Avila and Bruce J. Rounsaville, "Opioid-, Cannabis- and Alcohol-Dependent Women Show More Rapid Progression to Substance Abuse Treatment," *Drug and Alcohol Dependence* 74, no. 3 (2004): 265–72.

145 **relatively few pot abusers seek treatment** Sharaf S. Khan et al., "Gender Differences in Cannabis Use Disorders: Results from the National Epidemiologic Survey of Alcohol and Related Conditions," *Drug and Alcohol Dependence* 130 (2013).

145 **women reported better-quality highs** Suzette M. Evans and Richard W. Foltin, "Exogenous Progesterone Attenuates the Subjective Effects of Smoked Cocaine in Women, but Not in Men," *Neuropsychopharmacology* 31, no. 3 (2006).

145 **female brains seem to suffer less damage** National Institute on Drug Abuse, "Gender Differences in Drug Abuse Risks and Treatment," September 1, 2000, https://archives.drugabuse.gov/news-events /nida-notes/2000/09/gender-differences-in-drug-abuse-risks -treatment.

145 **Nearly 13 percent of teen girls admit to using diet pills** National Institute on Drug Abuse, "Substance Use in Women," accessed February 28, 2020, https://www.drugabuse.gov/publications/research -reports/substance-use-in-women/summary.

146 **"intense spice, honey and dried orange peel aromas"** *Wine Spectator* review of Château d'Yquem Sauternes 1990, accessed February 28, 2020, https://www.wine.com/product/chateau-dyquem -sauternes-1990/7865.

Chapter 7: We Have to Talk About It: Starting the Conversation

155 **That percentage rises to 61 percent** Center on Addiction, "Teen Insights into Drugs, Alcohol and Nicotine: A National Survey of Adolescent Attitudes Toward Addictive Substances," June 2019, https://www.centeronaddiction.org/addiction-research/reports/teen -insights-drugs-alcohol-and-nicotine-national-survey-adolescent.

156 **they are significantly more likely to do so** Ibid.

164 **alcohol played some role in the plot** Hugh Klein and Kenneth S. Shiffman, "Alcohol-Related Content of Animated Cartoons: A Historical Perspective," *Frontiers in Public Health* 1 (2013).

164 **branded use of alcohol** James D. Sargent et al., "Alcohol Use in Motion Pictures and Its Relation with Early-Onset Teen Drinking," *Journal of Studies on Alcohol* 67, no. 1 (2006): 54–65.

164 **by an actor** Sonya Dal Cin et al., "Youth Exposure to Alcohol Use and Brand Appearances in Popular and Contemporary Movies," *Addiction* 103, no. 12 (2008): 1925–32.

164 **integrate the experience** Kurt Badenhausen, "Anheuser-Busch Launches Revolutionary Incentive-Based Sponsorship Model," *Forbes,* April 2, 2018, https://www.forbes.com/sites/kurtbaden

hausen/2018/04/02/anheuser-busch-launches-revolutionary
-incentive-based-sponsorship-model/#63f0c0cd3d5f.

165   **including smoking, binge drinking, and unprotected sex** Josh Compton and Elizabeth A. Craig, "Family Communication Patterns, Inoculation Theory, and Adolescent Substance-Abuse Prevention: Harnessing Post-Inoculation Talk and Family Communication Environments to Spread Positive Influence," *Journal of Family Theory & Review* (2019).

165   **not mentioned in the messaging** Kimberly A. Parker et al., "Inoculation's Efficacy with Young Adults' Risky Behaviors: Can Inoculation Confer Cross-Protection over Related but Untreated Issues," *Health Communication* 27, no. 3 (2012): 223–33.

175   **"they can give up their social lives and stay home with us"** Lisa Damour, *Untangled: Guiding Teenage Girls Through the Seven Transitions into Adulthood* (New York: Ballantine, 2016), 252.

175   **"I'm just interested in siding"** Ibid., 259.

Chapter 8: Everyone's Doing It: Friendship, Peer Pressure, and Substance Abuse

182   **"tough love" boot camp programs** Maia Szalavitz, *Help at Any Cost: How the Troubled-Teen Industry Cons Parents and Hurts Kids* (New York: Riverhead Books, 2006).

183   **"Peer use of substances has consistently"** David J. Hawkins and Richard F. Catalano, "Risk and Protective Factors for Alcohol and Other Drug Problems in Adolescence and Early Adulthood: Implications for Substance Abuse Prevention," *Psychological Bulletin* 112, no. 1 (1992).

189   **"For reasons not yet understood"** Margo Gardner and Laurence Steinberg, "Peer Influence on Risk Taking, Risk Preference, and Risky Decision Making in Adolescence and Adulthood: An Experimental Study," *Developmental Psychology* 41, no. 4 (2005).

190   **more than the negative risks** Laurence Steinberg, *Age of Opportunity: Lessons from the New Science of Adolescence* (New York: Mariner Books, 2015).

193   **"their friends would *not* condone frequent drug use"** Kelly L. Rulison, Megan E. Patrick, and Jennifer L. Maggs, "Linking Peer Relationships to Substance Use Across Adolescence," in *The Oxford*

Handbook of Adolescent Substance Abuse, ed. Robert A. Zucker and Sandra A. Brown (New York: Oxford University Press, 2019), 390–91.

196   **"you are really sharing parenting"** Frances E. Jensen and Amy Ellis Nutt, *The Teenage Brain: A Neuroscientist's Survival Guide to Raising Adolescents and Young Adults* (New York: Harper, 2016), 125.

196   **more likely to adapt to that perceived standard** Nicole J. Roberts and Diana Fishbein, "An Integrative Perspective on the Etiology of Substance Use," in *Prevention of Substance Use*, ed. Zili Sloboda, Hanno Petras, Elizabeth Robertson, and Ralph Hingson (New York: Springer, 2019), 45.

197   **affecting a little over 7 percent of people in one German survey** "Prevalence of Wine Intolerance: Results of a Survey from Mainz, Germany," *Deutsches Ärzteblatt International* 109, no. 25 (2012).

Chapter 9: The ABC's of Addiction Prevention: Best Practices for Schools

208   **"'The opposite of addiction'"** Judith Grisel, *Never Enough* (New York: Doubleday, 2019).

210   **inspecting and vaccinating children enrolled in public schools** Diane DeMuth Allensworth, "Health Services and Health Education," in *Prevention Science in School Settings: Complex Relationships and Processes*, ed. Kris Bosworth (New York: Springer, 2015), 107.

210   **illnesses get more regular care, and immunization rates go up** Ibid.

211   **Of those 57 percent, only 10 percent** Chris Ringwalt et al., "The Prevalence of Effective Substance Use Prevention Curricula in the Nation's High Schools," *Journal of Primary Prevention* 29, no. 6 (2008): 479–88.

212   **"maintain positive relationships, and make responsible decisions"** Jessica Newman and Linda Dusenbury, "Social and Emotional Learning (SEL): A Framework for Academic, Social, and Emotional Success," in *Prevention Science in School Settings: Complex Relationships and Processes*, ed. Kris Bosworth (New York: Springer, 2015), 289.

214   **or most other topics, for that matter** Scott Freeman et al., "Active Learning Increases Student Performance in Science, Engineering, and Mathematics," *Proceedings of the National Academy of Sciences* 11, no. 23 (2014): 8410–15.

214   peer-led discussions, small group activities, and role-playing Nancy
       Fictman Dana and Angela Hooser, "Teachers on the Front Line
       of Prevention Science," in *Prevention Science in School Settings:
       Complex Relationships and Processes*, ed. Kris Bosworth (New York:
       Springer, 2015), 89.

214   substance use, unplanned pregnancy, violence, depression, suicide
       Kris Bosworth, "Exploring the Intersection of Schooling and Pre-
       vention Science," in *Prevention Science in School Settings: Complex
       Relationships and Processes*, ed. Kris Bosworth (New York: Springer,
       2015), 3.

214   higher levels of emotional well-being Clea A. McNeely and
       James M. Nonnemaker, "Promoting School Connectedness: Evi-
       dence from the National Longitudinal Study of Adolescent Health,"
       Journal of School Health 72, no. 4 (2002): 138–40.

215   in order to become licensed as school counselors Amy Nitza, Ker-
       rie R. Fineran, and Brian Dobias, "Professional Counselors' Impact
       on Schools," in *Prevention Science in School Settings: Complex Relation-
       ships and Processes*, ed. Kris Bosworth (New York: Springer, 2015), 69.

216   can save communities as much as $38 Amounts vary based on pro-
       gram. For example, the Nurse-Family Partnership was found to
       save $2.88 for every dollar invested, Durham Connects saves $3.02
       for every dollar invested, the Good Behavior Game saves $25.92
       for every dollar invested, and LifeSkills Training saves $38.00 for
       every dollar invested.

218   improve the lives of the people they come in contact with Michele
       Borba, *Unselfie: Why Empathetic Kids Succeed in Our All-About-Me
       World* (New York: Touchstone, 2016).

218   "an 88 percent drop in 'proactive aggression'" Ibid.

Chapter 10: Healthily Ever After: Preventing Addiction in College and Beyond

237   "Students must be comfortable" Deborah A. Prentice and Dale T.
       Miller, "Pluralistic Ignorance and Alcohol Use on Campus: Some
       Consequences of Misperceiving the Social Norm," *Journal of Per-
       sonality and Social Psychology* 64, no. 2 (1993): 244.

238   The authors' survey found Ibid., 251.

239   "If the lines do coincide" Snopes.com, accessed March 1, 2020,
       https://www.snopes.com/fact-check/solo-cup-markings/.

239   **heaviest binge drinkers chug 72 percent** Henry Wechsler, "Binge Drinking on America's College Campuses: Findings from the Harvard School of Public Health College Alcohol Study," accessed March 1, 2020, https://popcenter.asu.edu/sites/default/files/prob lems/rape/PDFs/cas_mono_2000.pdf.

240   **"poor academic performance and college dropout rates"** Ibid.

240   **"only 3 percent of non-bingeing female students"** Ibid.

240   **drink more often, in greater amounts** John S. Baer, "Student Factors: Understanding Individual Variation in College Drinking," *Journal of Studies on Alcohol* 14 (2002): 40–53.

241   **activities such as music or art, and students who volunteer** Wechsler, "Binge Drinking on America's College Campuses."

241   **and City University of New York–Baruch College** *Princeton Review*, "Stone Cold Sober Schools," accessed March 1, 2020, https://www .princetonreview.com/college-rankings?rankings=stone-cold-sober -schools.

241   **and face more negative consequences** Baer, "Student Factors."

242   **most were able to walk away** Lee N. Robins et al., "How Permanent Was Vietnam Drug Addiction," *American Journal of Public Health* 64 (1974): 38–43.

242   **does not tend to persist after graduation** Kenneth J. Sher et al., "Personality and Substance Use Disorders: A Prospective Study," *Journal of Consulting and Clinical Psychology* 68, no. 5 (2000): 818–29.

243   **so we'd better know what we are talking about** Amanda M. Vader et al., "Where Do College Students Get Health Information? Believability and Use of Health Information Sources," *Health Promotion Practice* 12, no. 5 (2011): 713–22.

247   **has been associated with higher drinking levels** Baer, "Student Factors."

251   **"Drinking U"** CollegeStats.org, accessed March 1, 2020, https:// collegestats.org/explore/drinking-u/.

251   **Flip Cup rules at Rutgers and Yale** CollegeStats.org, accessed March 1, 2020, https://collegestats.org/explore/drinking-u/.

252   **five people drink in unison** *Skiing*, November 2, 2012, accessed March 1, 2020, https://www.skimag.com/videos/how-to-drink-a -shot-ski.

BIBLIOGRAPHY

Abar, Caitlin C. "Examining the Relationship Between Parenting Types and Patterns of Student Alcohol-Related Behavior During the Transition to College." *Psychology of Addictive Behaviors* 26, no. 1 (2012): 20.

Abel, E. L. *Marihuana.* Springer Science & Business Media, 2013.

Adlaf, Edward M., Frank J. Ivis, Reginald G. Smart, and Gordon W. Walsh. "Enduring Resurgence or Statistical Blip? Recent Trends from the Ontario Student Drug Use Survey." *Canadian Journal of Public Health/Revue canadienne de sante publique* 87, no. 3 (1996): 189–92.

Aiken, Mary. *The Cyber Effect: An Expert in Cyberpsychology Explains How Technology Is Shaping Our Children, Our Behavior, and Our Values— and What We Can Do About It.* Spiegel & Grau, 2017.

Alter, Adam. *Irresistible: The Rise of Addictive Technology and the Business of Keeping Us Hooked.* Penguin Books, 2018.

Amato, Paul R. "Children of Divorce in the 1990s: An Update of the Amato and Keith (1991) Meta-Analysis." *Journal of Family Psychology* 15, no. 3 (2001): 355.

American Medical Association. *Nostrums and Quackery: Articles on the Nostrum Evil and Quackery Reprinted from the Journal of the American Medical Association.* Press of American Medical Association, 1921.

American Psychological Association. "Stress in America: Generation Z." *Stress in America Survey,* January 14 (2018): 2019.

Argyriou, Evangelia, Miji Um, Claire Carron, and Melissa A. Cyders. "Age and Impulsive Behavior in Drug Addiction: A Review of Past Research and Future Directions." *Pharmacology Biochemistry and Behavior* 164 (2018): 106–17.

Arkes, Jeremy. "The Temporal Effects of Parental Divorce on Youth Substance Use." *Substance Use & Misuse* 48, no. 3 (2013): 290–97.

Armstrong, Thomas. *The Power of the Adolescent Brain: Strategies for Teaching Middle and High School Students*. ASCD, 2016.

Armstrong, Tonya D., and E. Jane Costello. "Community Studies on Adolescent Substance Use, Abuse, or Dependence and Psychiatric Comorbidity." *Journal of Consulting and Clinical Psychology* 70, no. 6 (2002): 1224.

Arnold, James N. *The Narragansett Historical Register (R1)*. Heritage Books, 1996.

Arria, Amelia M., Kimberly M. Caldeira, Hannah K. Allen, Brittany A. Bugbee, Kathryn B. Vincent, and Kevin E. O'Grady. "Prevalence and Incidence of Drug Use Among College Students: An 8-Year Longitudinal Analysis." *American Journal of Drug and Alcohol Abuse* 43, no. 6 (2017): 711–18.

Asprey, Dave. *Head Strong: The Bulletproof Plan to Activate Untapped Brain Energy to Work Smarter and Think Faster—in Just Two Weeks*. Harper Wave, 2017.

Atkin, Charles, John Hocking, and Martin Block. "Teenage Drinking: Does Advertising Make a Difference." *Journal of Communication* 34, no. 2 (1984): 157–67.

Ausstin, Erica Weintraub, and Bethe Nach-Ferguson. "Sources and Influences of Young School-Age Children's General and Brand-Specific Knowledge About Alcohol." *Health Communication* 7, no. 1 (1995): 1–20.

Baer, John S. "Student Factors: Understanding Individual Variation in College Drinking." *Journal of Studies on Alcohol*, Supplement 14 (2002): 40–53.

Bandura, Albert. *Self-Efficacy: The Exercise of Control*. Worth Publishers, 1997.

Barnett, Robin. *Addict in the House: A No-Nonsense Family Guide Through Addiction and Recovery*. New Harbinger Publications, 2016.

Barres, Romain, Jie Yan, Brendan Egan, Jonas Thue Treebak, Morten Rasmussen, Tomas Fritz, Kenneth Caidahl, Anna Krook, Donal J. O'Gorman, and Juleen R Zierath. "Acute Exercise Remodels Promoter Methylation in Human Skeletal Muscle." *Cell Metabolism* 15, no. 3 (2012): 405–11.

Barrett, Anne E., and R. Jay Turner. "Family Structure and Mental Health: The Mediating Effects of Socioeconomic Status, Family Process, and

Social Stress." *Journal of Health and Social Behavior* 46, no. 2 (2005): 156–69.

Baum, Dan. "Legalize It All." *Harper's Magazine* 24 (2016).

Baumrind, Diana. "The Influence of Parenting Style on Adolescent Competence and Substance Use." *Journal of Early Adolescence* 11, no. 1 (1991): 56–95.

Beauvais, Fred. "American Indians and Alcohol." *Alcohol Research* 22, no. 4 (1998): 253.

Beetz, Andrea, Kerstin Uvnäs-Moberg, Henri Julius, and Kurt Kotrschal. "Psychosocial and Psychophysiological Effects of Human-Animal Interactions: The Possible Role of Oxytocin." *Frontiers in Psychology* 3 (2012): 234.

Begun, Audrey, Diana DiNitto, and Shulamith Lala A. Straussner. *Implementing the Grand Challenge of Reducing and Preventing Alcohol Misuse and Its Consequences.* Routledge, 2018.

Bercaw, Nancy. *Dryland.* Grand Harbor Press, 2017.

Berget, Bente, Øivind Ekeberg, and Bjarne O Braastad. "Animal-Assisted Therapy with Farm Animals for Persons with Psychiatric Disorders: Effects on Self-Efficacy, Coping Ability and Quality of Life, a Randomized Controlled Trial." *Clinical Practice and Epidemiology in Mental Health* 4, no. 1 (2008): 9.

Bhatta, Dharma N., and Stanton A. Glantz. "Association of E-Cigarette Use with Respiratory Disease Among Adults: A Longitudinal Analysis." *American Journal of Preventive Medicine* (2019).

Bisaga, Adam, and Karen Chernyaev. *Overcoming Opioid Addiction: The Authoritative Medical Guide for Patients, Families, Doctors, and Therapists.* The Experiment, 2018.

Black, Claudia. *Unspoken Legacy: Addressing the Impact of Trauma and Addiction Within the Family.* Central Recovery Press, 2018.

Borba, Michele. *The Big Book of Parenting Solutions.* John Wiley & Sons, 2009.
———. *Unselfie: Why Empathetic Kids Succeed in Our All-About-Me World.* Simon & Schuster, 2016.

Borsari, Brian, and Kate B. Carey. "Descriptive and Injunctive Norms in College Drinking: A Meta-Analytic Integration." *Journal of Studies on Alcohol* 64, no. 3 (2003): 331–41.

Bosworth, Kris. *Prevention Science in School Settings: Complex Relationships and Processes.* Springer, 2015.

Botvin, Gilbert J., and Kenneth W. Griffin. "Life Skills Training: Preventing

Substance Misuse by Enhancing Individual and Social Competence." *New Directions for Youth Development*, no. 141 (2014): 57–65.

Bourgois, Philippe, and Jeffrey Schonberg. *Righteous Dopefiend*. University of California Press, 2009.

Boyd, J. Wesley, and Eric Metcalf. *Almost Addicted: Is My (or My Loved One's) Drug Use a Problem? (the Almost Effect)*. Hazelden, 2012.

Brand, Evan. *The Everything Guide to Nootropics: Boost Your Brain Function with Smart Drugs and Memory Supplements*. Everything, 2016.

Brand, Russell. *Recovery*. Henry Holt, 2017.

Brecht, Mary-Lynn, Ann O'Brien, Christina Von Mayrhauser, and M. Douglas Anglin. "Methamphetamine Use Behaviors and Gender Differences." *Addictive Behaviors* 29, no. 1 (2004): 89–106.

Brewer, Judson. *The Craving Mind: From Cigarettes to Smartphones to Love—Why We Get Hooked and How We Can Break Bad Habits*. Yale University Press, 2018.

Brooks, Robert, and Sam Goldstein. *Nurturing Resilience in Our Children*. McGraw-Hill Professional, 2002.

———. *Raising Resilient Children: Fostering Strength, Hope, and Optimism in Your Child*. McGraw-Hill Education, 2002.

Brown, Sandra A., Susan F. Tapert, Eric Granholm, and Dean C. Delis. "Neurocognitive Functioning of Adolescents: Effects of Protracted Alcohol Use." *Alcoholism: Clinical and Experimental Research* 24, no. 2 (2000): 164–71.

Burke-Harris, Nadine. *The Deepest Well: Healing the Long-Term Effects of Childhood Adversity*. Houghton Mifflin Harcourt, 2018.

Burnham, John C. *Bad Habits*. NYU Press, 1993.

Burns, Eric. *Spirits of America: A Social History of Alcohol*. Temple University Press, 2010.

Burton, R., and N. Sheron. "No Level of Alcohol Consumption Improves Health." *Lancet* (2018).

Butler, Jon. *Becoming America: The Revolution Before 1776*. Harvard University Press, 2001.

Califano, Joseph A., Jr. *How to Raise a Drug-Free Kid: The Straight Dope for Parents*. Atria Books, 2014.

Carey, Benedict. *How We Learn: The Surprising Truth About When, Where, and Why It Happens*. Random House Trade Paperbacks, 2015.

Carskadon, Mary A. "Sleep in Adolescents: The Perfect Storm." *Pediatric Clinics* 58, no. 3 (2011): 637–47.

Castle, David J. "Cannabis and Psychosis: What Causes What." *F1000 Medicine Reports* 5 (2013).

Centers for Disease Control and Prevention. "Outbreak of Lung Injury Associated with the Use of E-Cigarette, or Vaping, Products." *What Is New* (2019).

Cheever, Susan. *As Good as I Could Be.* Simon & Schuster, 2001.

———. *Drinking in America: Our Secret History.* Twelve, 2016.

———. *Home Before Dark.* Bantam, 1991.

———. *Note Found in a Bottle (Wsp Readers Club).* Washington Square Press, 2000.

———. *Treetops: A Family Memoir.* Bantam, 1991.

Clegg, Bill. *Portrait of an Addict as a Young Man.* Back Bay Books, 2010.

Compton, Josh, and Elizabeth A. Craig. "Family Communication Patterns, Inoculation Theory, and Adolescent Substance-abuse Prevention: Harnessing Post-inoculation Talk and Family Communication Environments to Spread Positive Influence." *Journal of Family Theory & Review* (2019).

Conover, Marilyn. *Hannah Jumper.* Meristem. 1994.

Cozolino, Louis. *The Neuroscience of Human Relationships: Attachment and the Developing Social Brain.* 2nd ed., Norton Series on Interpersonal Neurobiology. W. W. Norton & Company, 2014.

———. *The Social Neuroscience of Education: Optimizing Attachment and Learning in the Classroom.* Norton Series on the Social Neuroscience of Education. W. W. Norton & Company, 2013.

Crandall, Mark. *Eulogy of Childhood Memories.* 2017.

Creagh, Milton L. *Nobody Wants Your Child.* Rock Hill Books of Georgia, 2006.

Crome, Ilana, and Richard Williams. *Substance Misuse and Young People.* CRC Press, 2019.

Cservenka, Anita, and Ty Brumback. "The Burden of Binge and Heavy Drinking on the Brain: Effects on Adolescent and Young Adult Neural Structure and Function." *Frontiers in Psychology* 8 (2017): 1111.

Cutt, Hayley, Billie Giles-Corti, Matthew Knuiman, Anna Timperio, and Fiona Bull. "Understanding Dog Owners' Increased Levels of Physical Activity: Results from Reside." *American Journal of Public Health* 98, no. 1 (2008): 66–69.

Cyders, M. A., T. C. B. Zapolski, and J. L. Combs. "Experimental Effect of Positive Urgency on Negative Outcomes from Risk Taking and on Increased Alcohol Consumption." *Psychology of Addictive Behaviors* 24, no. 3 (2010): 367–75.

Cyders, Melissa A., Kate Flory, Sarah Rainer, and Gregory T. Smith. "The Role of Personality Dispositions to Risky Behavior in Predicting First-Year College Drinking." *Addiction* 104, no. 2 (2009): 193–202.

Dal Cin, Sonya, Keilah A. Worth, Madeline A. Dalton, and James D. Sargent. "Youth Exposure to Alcohol Use and Brand Appearances in Popular Contemporary Movies." *Addiction* 103, no. 12 (2008): 1925–32.

Damasio, Antonio. *Self Comes to Mind: Constructing the Conscious Brain.* Vintage, 2012.

Damon, William. *The Path to Purpose: How Young People Find Their Calling in Life.* Free Press, 2009.

Damour, Lisa. *Under Pressure.* Ballantine Books, 2019.

———. *Untangled.* Ballantine Books, 2016.

Day, Jeremy J., and Regina M. Carelli. "The Nucleus Accumbens and Pavlovian Reward Learning." *Neuroscientist* 13, no. 2 (2007): 148–59.

De Micheli, Denise, and Maria Lucia O. S. Formigoni. "Are Reasons for the First Use of Drugs and Family Circumstances Predictors of Future Use Patterns." *Addictive Behaviors* 27, no. 1 (2002): 87–100.

Deak, JoAnn, and Terrence Deak. *The Owner's Manual for Driving Your Adolescent Brain.* Little Pickle Press, 2013.

Dikel, William. *The Teacher's Guide to Student Mental Health.* W. W. Norton & Company, 2014.

Dluzen, Dean E., and Bin Liu. "Gender Differences in Methamphetamine Use and Responses: A Review." *Gender Medicine* 5, no. 1 (2008): 24–35.

Dodes, Lance, and Zachary Dodes. *The Sober Truth: Debunking the Bad Science Behind 12-Step Programs and the Rehab Industry.* Beacon Press, 2015.

Dodge, Tonya L., and James J. Jaccard. "The Effect of High School Sports Participation on the Use of Performance-Enhancing Substances in Young Adulthood." *Journal of Adolescent Health* 39, no. 3 (2006): 367–73.

Doidge, Norman. *The Brain That Changes Itself.* Scribe Publications, 2010.

———. *The Brain's Way of Healing: Remarkable Discoveries and Recoveries from the Frontiers of Neuroplasticity.* Penguin Books, 2016.

Donovan, John E. "Really Underage Drinkers: The Epidemiology of Children's Alcohol Use in the United States." *Prevention Science* 8, no. 3 (2007): 192.

Duell, N., and L. Steinberg. "Positive Risk Taking in Adolescence." *Child Development Perspectives* (2019).

Dumas, Tara M., Wendy E. Ellis, and David A. Wolfe. "Identity Development as a Buffer of Adolescent Risk Behaviors in the Context of Peer Group Pressure and Control." *Journal of Adolescence* 35, no. 4 (2012): 917–27.

Edens, Anette. *From Monsters to Miracles: Parent Driven Recovery Tools That Work*. Lulu Publishing Services, 2016.

Ekamper, Peter, F. Van Poppel, A. D. Stein, and L. H. Lumey. "Independent and Additive Association of Prenatal Famine Exposure and Intermediary Life Conditions with Adult Mortality Between Age 18–63 Years." *Social Science & Medicine* 119 (2014): 232–39.

Engle, Patrice L., and Maureen M. Black. "The Effect of Poverty on Child Development and Educational Outcomes." *Annals of the New York Academy of Sciences* 1136, no. 1 (2008): 243–56.

Erickson, Carlton K. *The Science of Addiction: From Neurobiology to Treatment*. W. W. Norton & Company, 2007.

Erikson, Erik H., and Robert Coles. *The Erik Erikson Reader*. W. W. Norton & Company, 2000.

"Etymonline.com." https://www.etymonline.com/word/addict.

Evans, Suzette M., and Richard W. Foltin. "Exogenous Progesterone Attenuates the Subjective Effects of Smoked Cocaine in Women, but Not in Men." *Neuropsychopharmacology* 31, no. 3 (2006): 659.

Ewald, D. Rose, Robert W. Strack, and Muhsin Michael Orsini. "Rethinking Addiction." *Global Pediatric Health* 6 (2019): 2333794X18821943.

Eyler, Joshua R. *How Humans Learn: The Science and Stories Behind Effective College Teaching*. West Virginia University Press, 2018.

Faber, Adele, and Elaine Mazlish. *How to Talk So Kids Will Listen & Listen So Kids Will Talk*. Simon & Schuster, 2012.

Fagell, Phyllis L. *Middle School Matters: The 10 Key Skills Kids Need to Thrive in Middle School and Beyond—and How Parents Can Help*. Da Capo Lifelong Books, 2019.

Felitti, Vincent J. "The Origins of Addiction: Evidence from the Adverse Childhood Experiences Study." (2004).

Finan, Christopher M. *Drunks: An American History*. Beacon Press, 2017.

Finkelhor, David, Anne Shattuck, Heather A. Turner, and Sherry L. Hamby. "The Lifetime Prevalence of Child Sexual Abuse and Sexual

Assault Assessed in Late Adolescence." *Journal of Adolescent Health* 55, no. 3 (2014): 329–33.

Finnigan, Candy, and Sean Finnigan. *When Enough Is Enough: A Comprehensive Guide to Successful Intervention.* Avery, 2008.

Fletcher, Anne M. *Inside Rehab: The Surprising Truth About Addiction Treatment—and How to Get Help That Works.* Penguin Books, 2013.

Flores, Philip J. *Addiction as an Attachment Disorder.* Jason Aronson Inc., 2011.

Foote, Jeffrey, Carrie Wilkens, Nicole Kosanke, and Stephanie Higgs. *Beyond Addiction: How Science and Kindness Help People Change.* Scribner, 2014.

Frances, Allen. *Saving Normal: An Insider's Revolt Against Out-of-control Psychiatric Diagnosis,* DSM-5, *Big Pharma, and the Medicalization of Ordinary Life.* William Morrow Paperbacks, 2014.

Freeman, Scott, Sarah L. Eddy, Miles McDonough, Michelle K. Smith, Nnadozie Okoroafor, Hannah Jordt, and Mary Pat Wenderoth. "Active Learning Increases Student Performance in Science, Engineering, and Mathematics." *Proceedings of the National Academy of Sciences* 111, no. 23 (2014): 8410–15.

Frijns, Tom, Catrin Finkenauer, and Loes Keijsers. "Shared Secrets Versus Secrets Kept Private Are Linked to Better Adolescent Adjustment." *Journal of Adolescence* 36, no. 1 (2013): 55–64.

Froiland, John Mark, and Noelle Whitney. "Parenting Style, Gender, Beer Drinking and Drinking Problems of College Students." *International Journal of Psychology: A Biopsychosocial Approach* 16 (2015): 93–109.

Furnas, J. C. *The Americans: A Social History of the United States, 1587–1914.* Putnam, 1969.

Gaine, Marie E., Snehajyoti Chatterjee, and Ted Abel. "Sleep Deprivation and the Epigenome." *Frontiers in Neural Circuits* 12 (2018): 14.

Gardner, Margo, and Laurence Steinberg. "Peer Influence on Risk Taking, Risk Preference, and Risky Decision Making in Adolescence and Adulthood: An Experimental Study." *Developmental Psychology* 41, no. 4 (2005): 625.

Garmezy, Norman, and Michael Rutter. *Stress, Coping, and Development in Children.* Johns Hopkins University Press, 1988.

Gilchrist, Lewayne D., and Steven Paul Schinke. *Preventing Social and Health Problems Through Life Skills Training.* University of Washington, 1985.

Ginsburg, Kenneth R. *Building Resilience in Children and Teens: Giving Kids Roots and Wings*. American Academy of Pediatrics, 2014.

———. *Raising Kids to Thrive: Balancing Love with Expectations and Protection with Trust*. American Academy of Pediatrics, 2015.

Glantz, Stanton A., and David W Bareham. "E-Cigarettes: Use, Effects on Smoking, Risks, and Policy Implications." *Annual Review of Public Health* 39 (2018): 215–35.

Gnaulati, Enrico. *Back to Normal: Why Ordinary Childhood Behavior Is Mistaken for ADHD, Bipolar Disorder, and Autism Spectrum Disorder*. Beacon Press, 2014.

Gold, Claudia M. *The Silenced Child: From Labels, Medications, and Quick-Fix Solutions to Listening, Growth, and Lifelong Resilience*. Da Capo Lifelong Books, 2016.

Goleman, Daniel. *Focus: The Hidden Driver of Excellence*. Harper Paperbacks, 2015.

Gradisar, Michael, Amy R. Wolfson, Allison G. Harvey, Lauren Hale, Russell Rosenberg, and Charles A. Czeisler. "The Sleep and Technology Use of Americans: Findings from the National Sleep Foundation's 2011 Sleep in America Poll." *Journal of Clinical Sleep Medicine* 9, no. 12 (2013): 1291–99.

Graham, James. *The Secret History of Alcoholism*. Element Books Limited, 1996.

Grant, Jon E., and Samuel R. Chamberlain. "Expanding the Definition of Addiction: DSM-5 Vs. ICD-11." *CNS Spectrums* 21, no. 4 (2016): 300–303.

Grasse, Steven. *Colonial Spirits: A Toast to Our Drunken History*. Harry N. Abrams, 2016.

Grim, Ryan. *This Is Your Country on Drugs: The Secret History of Getting High in America*. Turner, 2010.

Grisel, Judith. *Never Enough*. Doubleday, 2019.

Griswold, Max G., Nancy Fullman, Caitlin Hawley, Nicholas Arian, Stephanie R. M. Zimsen, Hayley D. Tymeson, Vidhya Venkateswaran, Austin Douglas Tapp, Mohammad H. Forouzanfar, and Joseph S. Salama. "Alcohol Use and Burden for 195 Countries and Territories, 1990–2016: A Systematic Analysis for the Global Burden of Disease Study 2016." *Lancet* 392, no. 10152 (2018): 1015–35.

Grosso, Chris. *Dead Set on Living: Making the Difficult but Beautiful Journey from F#*king Up to Waking Up*. Gallery Books, 2018.

Guller, Leila, Tamika C. B. Zapolski, and Gregory T. Smith. "Personality Measured in Elementary School Predicts Middle School Addictive Behavior Involvement." *Journal of Psychopathology and Behavioral Assessment* 37, no. 3 (2015): 523–32.

Gunaratana, Bhante Henepola. *Mindfulness in Plain English*. Wisdom Publications, 2011.

Hadland, Scott E., John R. Knight, and Sion K. Harris. "Alcohol Use Disorder: A Pediatric-Onset Condition Needing Early Detection and Intervention." *Pediatrics* (2019).

Hampton, Ryan. *American Fix: Inside the Opioid Addiction Crisis—and How to End it*. All Points Books, 2018.

Hanchett, Janelle. *I'm Just Happy to Be Here*. Hachette Books, 2018.

Hanson, Rick. *Hardwiring Happiness: The New Brain Science of Contentment, Calm, and Confidence*. Harmony, 2016.

Hari, Johann. "The Likely Cause of Addiction Has Been Discovered, and It Is Not What You Think." *Huffpost* (2015).

———. *Lost Connections: Uncovering the Real Causes of Depression—and the Unexpected Solutions*. Bloomsbury USA, 2018.

Harris, Judith Rich. *The Nurture Assumption*. Simon & Schuster, 1999.

Hart, Carl. *High Price: A Neuroscientist's Journey of Self-Discovery That Challenges Everything You Know About Drugs and Society*. Harper Perennial, 2014.

Hartley-Brewer, Elizabeth. *Talking to Tweens*. Da Capo Press, 2009.

Hawkins, J. David, Richard F. Catalano, and Janet Y. Miller. "Risk and Protective Factors for Alcohol and Other Drug Problems in Adolescence and Early Adulthood: Implications for Substance Abuse Prevention." *Psychological Bulletin* 112, no. 1 (1992): 64.

Hayden, Brian, Neil Canuel, and Jennifer Shanse. "What Was Brewing in the Natufian? An Archaeological Assessment of Brewing Technology in the Epipaleolithic." *Journal of Archaeological Method and Theory* 20, no. 1 (2013): 102–50.

Heather, Nick, David Best, Anna Kawalek, Matt Field, Marc Lewis, Frederick Rotgers, Reinout W. Wiers, and Derek Heim. "Challenging the Brain Disease Model of Addiction: European Launch of the Addiction Theory Network." *Addiction Research and Theory* 26, no. 4 (2018).

Heilig, Markus. *The Thirteenth Step: Addiction in the Age of Brain Science*. Columbia University Press, 2015.

Heitner, Devorah. *Screenwise: Helping Kids Thrive (and Survive) in Their Digital World*. Routledge, 2016.

Hernandez-Avila, Carlos A., Bruce J. Rounsaville, and Henry R. Kranzler. "Opioid-, Cannabis- and Alcohol-Dependent Women Show More Rapid Progression to Substance Abuse Treatment." *Drug and Alcohol Dependence* 74, no. 3 (2004): 265–72.

Herren, Chris, and Bill Reynolds. *Basketball Junkie*. St. Martin's Press, 2011.

Hershberger, A. R., M. Um, and M. A. Cyders. "The Relationship Between the Upps-P Impulsive Personality Traits and Substance Use Psychotherapy Outcomes: A Meta-Analysis." *Drug and Alcohol Dependence* (2017).

Holland, Barbara. *The Joy of Drinking*. Bloomsbury Publishing USA, 2008.

Hoshi, Rosa, Hannah Pratt, Sachin Mehta, Alyson J. Bond, and H. Valerie Curran. "An Investigation into the Sub-Acute Effects of Ecstasy on Aggressive Interpretative Bias and Aggressive Mood—Are There Gender Differences." *Journal of Psychopharmacology* 20, no. 2 (2006): 291–301.

Gateley, Ian. *Drink: A Cultural History of Alcohol*. Gotham Books, 2008.

Immordino-Yang, Mary Helen. *Emotions, Learning, and the Brain: Exploring the Educational Implications of Affective Neuroscience*. W. W. Norton & Company, 2015.

Immordino-Yang, Mary Helen, Linda Darling-Hammond, and Christine Krone. "The Brain Basis for Integrated Social, Emotional, and Academic Development." Aspen Institute (2018).

Jacob, Joseph W. *Medical Uses of Marijuana*. Trafford Publishing, 2009.

Jacob, T., and S. L. Johnson. "Family Influences on Alcohol and Other Substance Use." *Sourcebook on Substance Abuse* (1999): 165–74.

Jamison, Leslie. *The Recovering*. Granta Books, 2018.

Jenkins, Tiffany. *High Achiever*. Harmony, 2019.

Jennison, Karen M., and Kenneth A. Johnson. "Alcohol Dependence in Adult Children of Alcoholics: Longitudinal Evidence of Early Risk." *Journal of Drug Education* 28, no. 1 (1998): 19–37.

Jensen, Frances E., and Amy Ellis Nutt. *The Teenage Brain: A Neuroscientist's Survival Guide to Raising Adolescents and Young Adults*. Harper Paperbacks, 2016.

Johnston, Ann Dowsett. *Drink*. HarperCollins, 2013.

Kahn, Jeffrey P. "How Beer Gave Us Civilization." *New York Times*, March 15, 2013.

———. *Angst*. Oxford University Press, 2012.

Kaminsky, Zachary, Arturas Petronis, Sun-Chong Wang, Brian Levine, Omar Ghaffar, Darlene Floden, and Anthony Feinstein. "Epigenetics of Personality Traits: An Illustrative Study of Identical Twins Discordant for Risk-Taking Behavior." *Twin Research and Human Genetics* 11, no. 1 (2008): 1–11.

Kamioka, Hiroharu, Shinpei Okada, Kiichiro Tsutani, Hyuntae Park, Hiroyasu Okuizumi, Shuichi Handa, Takuya Oshio, Sang-Jun Park, Jun Kitayuguchi, and Takafumi Abe. "Effectiveness of Animal-Assisted Therapy: A Systematic Review of Randomized Controlled Trials." *Complementary Therapies in Medicine* 22, no. 2 (2014): 371–90.

Kandall, Stephen R. *Substance and Shadow*. Harvard University Press, 1999.

Kandel, Denise, and Richard Faust. "Sequence and Stages in Patterns of Adolescent Drug Use." *Archives of General Psychiatry* 32, no. 7 (1975): 923–32.

Kandel, Denise, and Kazuo Yamaguchi. "From Beer to Crack: Developmental Patterns of Drug Involvement." *American Journal of Public Health* 83, no. 6 (1993): 851–55.

Kandel, Denise B., and Mark Davies. "High School Students Who Use Crack and Other Drugs." *Archives of General Psychiatry* 53, no. 1 (1996): 71–80.

Kardaras, Nicholas. *Glow Kids: How Screen Addiction Is Hijacking Our Kids—and How to Break the Trance*. St. Martin's Griffin, 2017.

Karr, Mary. *Lit*. HarperCollins, 2009.

Karr-Morse, Robin. *Scared Sick: The Role of Childhood Trauma in Adult Disease*. Basic Books, 2012.

Kastner, Laura S., and Jennifer Wyatt. *Getting to Calm: Cool-Headed Strategies for Parenting Tweens + Teens—Updated and Expanded*. Parent Map, 2018.

Katz, Mark. *On Playing a Poor Hand Well*. W. W. Norton & Company, 1997.

Kelly, Anita E. *The Psychology of Secrets*. Springer Science & Business Media, 2012.

Ketcham, Katherine. *The Only Life I Could Save*. Sounds True, 2018.

Ketcham, Katherine, William F. Asbury, Mel Schulstad, and Arthur P. Ciaramicoli. *Beyond the Influence: Understanding and Defeating Alcoholism*. Bantam, 2000.

Khan, Sharaf, Mayumi Okuda, Deborah S. Hasin, Roberto Secades-Villa, Katherine Keyes, Keng-Han Lin, Bridget Grant, and Carlos Blanco. "Gender Differences in Lifetime Alcohol Dependence: Results from the National Epidemiologic Survey on Alcohol and Related Conditions." *Alcoholism: Clinical and Experimental Research* 37, no. 10 (2013): 1696–1705.

Khan, Sharaf S., Roberto Secades-Villa, Mayumi Okuda, Shuai Wang, Gabriela Pérez-Fuentes, Bradley T. Kerridge, and Carlos Blanco. "Gender Differences in Cannabis Use Disorders: Results from the National Epidemiologic Survey of Alcohol and Related Conditions." *Drug and Alcohol Dependence* 130, no. 1–3 (2013): 101–8.

Khar, Erin. *Strung Out.* Harlequin, 2020.

Khurana, Atika, Daniel Romer, Laura M. Betancourt, Nancy L. Brodsky, Joan M. Giannetta, and Hallam Hurt. "Experimentation Versus Progression in Adolescent Drug Use: A Test of an Emerging Neurobehavioral Imbalance Model." *Development and Psychopathology* 27, no. 3 (2015): 901–13.

King, Heather. *Parched.* Penguin, 2006.

Kipper, David, and Stephen Whitney. *The Addiction Solution: Unraveling the Mysteries of Addiction Through Cutting-Edge Brain Science.* Audible Studios on Brilliance Audio, 2016.

Klein, Hugh, and Kenneth S. Shiffman. "Alcohol-Related Content of Animated Cartoons: A Historical Perspective." *Frontiers in Public Health* 1 (2013): 2.

Koss, Mary P., Nicole P. Yuan, Douglas Dightman, Ronald J. Prince, Mona Polacca, Byron Sanderson, and David Goldman. "Adverse Childhood Exposures and Alcohol Dependence Among Seven Native American Tribes." *American Journal of Preventive Medicine* 25, no. 3 (2003): 238–44.

Krasnostein, Sarah. *The Trauma Cleaner.* St. Martin's Press, 2018.

Krosoczka, Jarrett J. *Hey, Kiddo.* Scholastic, 2018.

Krovitz-Neren, Barbara. *Parenting the Addicted Teen: A 5-Step Foundational Program.* Central Recovery Press, 2017.

Kuhn, Cynthia, Scott Swartzwelder, and Wilkie Wilson. *Buzzed: The Straight Facts About the Most Used and Abused Drugs from Alcohol to Ecstasy (Fully Revised and Updated Fourth Edition).* W. W. Norton & Company, 2014.

Kuhn, Cynthia, Scott Swartzwelder, and Wilkie Wilson. *Buzzed: The

Straight Facts About the Most Used and Abused Drugs from Alcohol to Ecstasy, Fifth Edition. W. W. Norton & Company, 2019.

Kulig, Kimary, Nancy D. Brener, and Tim McManus. "Sexual Activity and Substance Use Among Adolescents by Category of Physical Activity Plus Team Sports Participation." *Archives of Pediatrics & Adolescent Medicine* 157, no. 9 (2003): 905–12.

Lamarine, Roland J. "A Pilot Study of Sources of Information and Substance Use Patterns Among Selected American Indian High School Seniors." *Journal of American Indian Education* (1993): 30–39.

Le Foll, Bernard, Alexandra Gallo, Yann Le Strat, Lin Lu, and Philip Gorwood. "Genetics of Dopamine Receptors and Drug Addiction: A Comprehensive Review." *Behavioural Pharmacology* 20, no. 1 (2009): 1–17.

Lee, Joseph. *Recovering My Kid: Parenting Young Adults in Treatment and Beyond.* Hazelden Publishing, 2012.

Lender, Mark Edward. *Drinking in America: A History.* Free Press, 1987.

Leukefeld, Carl G., and Thomas P. Gullotta. *Adolescent Substance Abuse.* Springer, 2018.

Levy, Sharon J. L., and Janet F. Williams. "Substance Use Screening, Brief Intervention, and Referral to Treatment." *Pediatrics* 138, no. 1 (2016): e20161211.

Lewis, Judith A., Robert Q. Dana, and Gregory A. Blevins. *Substance Abuse Counseling.* Cengage Learning, 2014.

Lewis, Katherine Reynolds. *The Good News About Bad Behavior: Why Kids Are Less Disciplined Than Ever—and What to Do About It.* PublicAffairs, 2018.

Lewis, Marc. "Addiction and the Brain: Development, Not Disease." *Neuroethics* 10, no. 1 (2017): 7–18.

———. *The Biology of Desire.* Scribe Publications, 2015.

———. *Memoirs of an Addicted Brain.* PublicAffairs, 2012.

Li, Yuanyuan, Qingyao Kong, Jiping Yue, Xuewen Gou, Ming Xu, and Xiaoyang Wu. "Genome-Edited Skin Epidermal Stem Cells Protect Mice from Cocaine-Seeking Behaviour and Cocaine Overdose." *Nature Biomedical Engineering* 3, no. 2 (2019): 105.

Logan, Ryan W., Brant P. Hasler, Erika E. Forbes, Peter L. Franzen, Mary M. Torregrossa, Yanhua H. Huang, Daniel J. Buysse, Duncan B. Clark, and Colleen A. McClung. "Impact of Sleep and Circadian

Rhythms on Addiction Vulnerability in Adolescents." *Biological Psychiatry* 83, no. 12 (2018): 987–96.

Lopez, Shane J. *Making Hope Happen: Create the Future You Want for Yourself and Others.* Simon & Schuster, 2014.

Luthar, Suniya S., Phillip J. Small, and Lucia Ciciolla. "Adolescents from Upper Middle Class Communities: Substance Misuse and Addiction Across Early Adulthood." *Development and Psychopathology* 30, no. 1 (2018): 315–35.

MacMillan, Thalia, and Amanda Sisselman-Borgia. *New Directions in Treatment, Education, and Outreach for Mental Health and Addiction.* Springer, 2018.

Macy, Beth. *Dopesick: Dealers, Doctors, and the Drug Company That Addicted America.* Little, Brown and Company, 2018.

Madden Ellsworth, Lindsay, Sarah Tragesser, and Ruth C. Newberry. "Interaction with Shelter Dogs Reduces Negative Affect of Adolescents in Substance Use Disorder Treatment." *Anthrozoös* 29, no. 2 (2016): 247–62.

Mancall, Peter C. *Deadly Medicine: Indians and Alcohol in Early America.* Cornell University Press, 1997.

Maric, Angelina, Eszter Montvai, Esther Werth, Matthias Storz, Janina Leemann, Sebastian Weissengruber, Christian C. Ruff, Reto Huber, Rositsa Poryazova, and Christian R. Baumann. "Insufficient Sleep: Enhanced Risk-seeking Relates to Low Local Sleep Intensity." *Annals of Neurology* 82, no. 3 (2017): 409–18.

Marlatt, G. Alan, Mary E. Larimer, and Katie Witkiewitz. *Harm Reduction, Second Edition: Pragmatic Strategies for Managing High-Risk Behaviors.* Guilford Press, 2011.

Masten, Ann, Vivian Faden, Robert Zucker, and Linda Spear. "Underage Drinking: A Developmental Framework." *Pediatrics* 121 (2008): S235–S251.

Maté, Gabor. *In the Realm of Hungry Ghosts: Close Encounters with Addiction.* North Atlantic Books, 2010.

———. *Scattered Minds: A New Look at the Origins and Healing of Attention Deficit Disorder.* Vintage Canada, 2000.

———. *When the Body Says No: Understanding the Stress-Disease Connection.* Wiley, 2011.

McCord, Joan, and William J. McCord. *Origins of Alcoholism.* Stanford University Press, 1960.

McKnight-Eily, Lela R., Danice K. Eaton, Richard Lowry, Janet B. Croft, Letitia Presley-Cantrell, and Geraldine S. Perry. "Relationships Between Hours of Sleep and Health-Risk Behaviors in US Adolescent Students." *Preventive Medicine* 53, no. 4–5 (2011): 271–73.

McNeely, Clea A., James M. Nonnemaker, and Robert W. Blum. "Promoting School Connectedness: Evidence from the National Longitudinal Study of Adolescent Health." *Journal of School Health* 72, no. 4 (2002): 138–46.

Medina, John. *Brain Rules (Updated and Expanded): 12 Principles for Surviving and Thriving at Work, Home, and School.* Pear Press, 2014.

Medina, Krista Lisdahl, Tim McQueeny, Bonnie J. Nagel, Karen L. Hanson, Tony T. Yang, and Susan F. Tapert. "Imaging Study: Prefrontal Cortex Morphometry in Abstinent Adolescent Marijuana Users: Subtle Gender Effects." *Addiction Biology* 14, no. 4 (2009): 457–68.

Meier, Barry. *Pain Killer: An Empire of Deceit and the Origin of America's Opioid Epidemic.* Random House, 2018.

Melnick, Merril J., Kathleen E. Miller, Donald F. Sabo, Michael P. Farrell, and Grace M. Barnes. "Tobacco Use Among High School Athletes and Nonathletes: Results of the 1997 Youth Risk Behavior Survey." *Adolescence* 36, no. 144 (2001): 727.

Milam, James R., and Katherine Ketcham. *Under the Infuence.* Madrona, 1981.

Miller, John G., and Karen G. Miller. *Raising Accountable Kids: How to Be an Outstanding Parent Using the Power of Personal Accountability.* TarcherPerigee, 2016.

Miller, William R., Alyssa A. Forcehimes, and Allen Zweben. *Treating Addiction: A Guide for Professionals.* Guilford Press, 2011.

Miner, Julianna. *Raising a Screen-Smart Kid: Embrace the Good and Avoid the Bad in the Digital Age.* TarcherPerigee, 2019.

Mitchell, Shannon Gwin, Jan Gryczynski, Arturo Gonzales, Ana Moseley, Thomas Peterson, Kevin E. O'Grady, and Robert P. Schwartz. "Screening, Brief Intervention, and Referral to Treatment (SBIRT) for Substance Use in a School-based Program: Services and Outcomes." *American Journal on Addictions* 21 (2012): S5–S13.

Mitchell, Tracey Helton. *The Big Fix.* Seal Press, 2016.

Moe, Jerry. *Understanding Addiction and Recovery Through a Child's Eyes: Hope, Help, and Healing for Families.* HCI, 2007.

Mohammad, Akikur. *The Anatomy of Addiction: What Science and Research Tell Us About the True Causes, Best Preventive Techniques, and Most Successful Treatments.* TarcherPerigee, 2016.

Monti, Peter M., Suzanne M. Colby, and Tracy O'Leary Tevyaw. *Adolescents, Alcohol, and Substance Abuse.* Guilford Press, 2012.

Moyers, William Cope, and Katherine Ketcham. *Broken.* Penguin, 2007.

Myers, Diane M., Brandi Simonsen, and George Sugai. "Increasing Teachers' Use of Praise with a Response-to-intervention Approach." *Education and Treatment of Children* 34, no. 1 (2011): 35–59.

Myers, Linda. *Stoney the Pony's Most Inspiring Year.* Inspiring Voices, 2012.

Naar, Sylvie, and Steven A. Safren. *Motivational Interviewing and CBT: Combining Strategies for Maximum Effectiveness (Applications of Motivational Interviewing).* Guilford Press, 2017.

Naar-King, Sylvie, and Mariann Suarez. *Motivational Interviewing with Adolescents and Young Adults (Applications of Motivational Interviewing).* Guilford Press, 2010.

Nagel, Bonnie J., Alecia D. Schweinsburg, Vinh Phan, and Susan F. Tapert. "Reduced Hippocampal Volume Among Adolescents with Alcohol Use Disorders Without Psychiatric Comorbidity." *Psychiatry Research: Neuroimaging* 139, no. 3 (2005): 181–90.

National Center on Addiction and Substance Abuse, Columbia University. *The Formative Years: Pathways to Substance Abuse Among Girls and Young Women Ages 8–22.* CASA, 2003.

National Center on the Humanities. "Becoming American: The British Atlantic Colonies, 1690–1763" (2009). http://nationalhumanities center.org/pds/becomingamer/ideas/text5/pennsylvaniagazette.pdf.

National Research Council. Division of Behavioral and Social Sciences and Education, Commission on Behavioral and Social Sciences and Education, and Committee on Substance Abuse Prevention Research. *Preventing Drug Abuse: What Do We Know.* National Academies Press, 1993.

National Research Council. Institute of Medicine, Board on Children, Youth, and Families, and Division of Behavioral and Social Sciences and Education. *Reducing Underage Drinking: A Collective Responsibility.* National Academies Press, 2004.

Neufeld, Gordon, and Gabor Maté. *Hold On to Your Kids: Why Parents Need to Matter More Than Peers.* Ballantine Books, 2006.

Nkansah-Amankra, Stephen, and Mark Minelli. "'Gateway Hypothesis' and Early Drug Use: Additional Findings from Tracking a Population-Based Sample of Adolescents to Adulthood." *Preventive Medicine Reports* 4 (2016): 134–41.

Nootropics Zone. "Nootropics: Unlocking Your True Potential with Smart Drugs" (2017): 117.

Obladen, Michael. "Lethal Lullabies: A History of Opium Use in Infants." *Journal of Human Lactation* 32, no. 1 (2016): 75–85.

O'Brien, Charles. "Addiction and Dependence in DSM-V." *Addiction* 106, no. 5 (2011): 866–67.

O'Brien, Charles P., Nora Volkow, and T. K. Li. "What's in a Word? Addiction Versus Dependence in DSM-V." (2006).

O'Connell, James J. *Stories from the Shadows: Reflections of a Street Doctor.* Boston Health Care for the Homeless, 2015.

Oettingen, Gabriele. *Rethinking Positive Thinking: Inside the New Science of Motivation.* Current, 2015.

One Hundred Years of Brewing—A Complete History of Progress Made in Art Science & Industry of Brewing in the World. Arno Press, 1974.

Osborn, Corinne O'Keefe. "A Guide to Addiction and Recovery for Native Americans." https://www.recovery.org/topics/native-americans/.

Pagliaro, Ann Marie, and Louis A. Pagliaro. *Substance Use Among Children and Adolescents: Its Nature, Extent, and Effects from Conception to Adulthood.* Wiley, 1996.

Palamar, Joseph J., Caroline Rutherford, and Katherine M. Keyes. "Summer as a Risk Factor for Drug Initiation." *Journal of General Internal Medicine* (2019): 1–3.

Pan, Wei, and Haiyan Bai. "A Multivariate Approach to a Meta-Analytic Review of the Effectiveness of the Dare Program." *International Journal of Environmental Research and Public Health* 6, no. 1 (2009): 267–77.

Parker, Kimberly A., Bobi Ivanov, and Josh Compton. "Inoculation's Efficacy with Young Adults' Risky Behaviors: Can Inoculation Confer Cross-Protection Over Related but Untreated Issues." *Health Communication* 27, no. 3 (2012): 223–33.

Patock-Peckham, Julie A., and Antonio A. Morgan-Lopez. "College Drinking Behaviors: Mediational Links Between Parenting Styles, Impulse Control, and Alcohol-Related Outcomes." *Psychology of Addictive Behaviors* 20, no. 2 (2006): 117.

Peele, Stanton. *Addiction Proof Your Child: A Realistic Approach to Preventing Drug, Alcohol, and Other Dependencies*. Harmony, 2007.

Perkins, H. Wesley. "The Emergence and Evolution of the Social Norms Approach to Substance Abuse Prevention." *The Social Norms Approach to Preventing School and College Age Substance Abuse: A Handbook for Educators, Counselors, and Clinicians* (2003): 3–17.

Perry, Bruce, and Maia Szalavitz. *The Boy Who Was Raised as a Dog*. Basic Books, 2007.

Perry, Jennifer L., Jane E. Joseph, Yang Jiang, Rick S. Zimmerman, Thomas H. Kelly, Mahesh Darna, Peter Huettl, Linda P. Dwoskin, and Michael T. Bardo. "Prefrontal Cortex and Drug Abuse Vulnerability: Translation to Prevention and Treatment Interventions." *Brain Research Reviews* 65, no. 2 (2011): 124–49.

Peterson, Christopher, Steven F. Maier, and Martin E. P. Seligman. *Learned Helplessness: A Theory for the Age of Personal Control*. Oxford University Press, 1995.

Petruzzello, Melissa. "7 of the World's Deadliest Plants." https://www.britannica.com/list/7-of-the-worlds-deadliest-plants.

Pfefferbaum, Adolf, Dongjin Kwon, Ty Brumback, Wesley K. Thompson, Kevin Cummins, Susan F. Tapert, Sandra A. Brown, Ian M. Colrain, Fiona C. Baker, and Devin Prouty. "Altered Brain Developmental Trajectories in Adolescents After Initiating Drinking." *American Journal of Psychiatry* 175, no. 4 (2017): 370–80.

Phelan, Thomas W. *Surviving Your Adolescents: How to Manage—and Let Go of—Your 13–18 Year Olds*. Parentmagic, 1998.

Phillips, Kaitlin Ugolik. *The Future of Feeling*. Little A, 2020.

Piper, Brian J., Christy L. Ogden, Olapeju M. Simoyan, Daniel Y. Chung, James F. Caggiano, Stephanie D. Nichols, and Kenneth L. McCall. "Trends in Use of Prescription Stimulants in the United States and Territories, 2006 to 2016." *PloS One* 13, no. 11 (2018): e0206100.

Poirier, Lionel A. "The Effects of Diet, Genetics and Chemicals on Toxicity and Aberrant DNA Methylation: An Introduction." *Journal of Nutrition* 132, no. 8 (2002): 2336S–9S.

Pollan, Michael. *The Botany of Desire: A Plant's-Eye View of the World*. Random House Trade Paperbacks, 2002.

———. *How to Change Your Mind*. Penguin Books, 2019.

Pooley, Clare. *The Sober Diaries*. Hachette UK, 2017.

Pope, Harrison G., Arthur Jacobs, Jean-Paul Mialet, Deborah Yurgelun-

Todd, and Staci Gruber. "Evidence for a Sex-Specific Residual Effect of Cannabis on Visuospatial Memory." *Psychotherapy and Psychosomatics* 66, no. 4 (1997): 179–84.

Powell, Douglas. *Teenagers: When to Worry and What to Do.* Main Street Books, 1987.

Prentice, Deborah A., and Dale T. Miller. "Pluralistic Ignorance and Alcohol Use on Campus: Some Consequences of Misperceiving the Social Norm." *Journal of Personality and Social Psychology* 64, no. 2 (1993): 243.

Presley, Cheryl A., Philip W. Meilman, and Jami S. Leichliter. "College Factors That Influence Drinking." *Journal of Studies on Alcohol,* Supplement 14 (2002): 82–90.

Quart, Alissa. *Hothouse Kids: The Dilemma of the Gifted Child.* Penguin Press, 2006.

Raab, Diana M., and James Brown. *Writers on the Edge: 22 Writers Speak About Addiction and Dependency.* Modern History Press, 2012.

Racal, Sarah Jane. "Social Perspectives of Addiction: Current Approaches and Underlying Challenges." *St. Theresa Journal of Humanities and Social Sciences* 2 (2016).

Raeburn, Paul. *Acquainted with the Night: A Parent's Quest to Understand Depression and Bipolar Disorder in His Children.* Broadway Books, 2005.

Raine, Adrian. *The Anatomy of Violence: The Biological Roots of Crime.* Vintage, 2014.

Rasmussen, Nicolas. *On Speed.* NYU Press, 2009.

Rawson, Richard A., Rachel Gonzales, Jeanne L. Obert, Michael J. McCann, and Paul Brethen. "Methamphetamine Use Among Treatment-Seeking Adolescents in Southern California: Participant Characteristics and Treatment Response." *Journal of Substance Abuse Treatment* 29, no. 2 (2005): 67–74.

Reckmeyer, Mary. *Strengths-Based Parenting: Developing Your Children's Innate Talents.* Gallup Press, 2016.

Reedy, Brad M. *The Journey of the Heroic Parent: Your Child's Struggle & the Road Home.* Regan Arts, 2016.

Rieder, Travis. *In Pain.* HarperCollins, 2019.

Ringwalt, Chris, Sean Hanley, Amy A. Vincus, Susan T. Ennett, Louise A. Rohrbach, and J. Michael Bowling. "The Prevalence of Effective Substance Use Prevention Curricula in the Nation's High Schools." *Journal of Primary Prevention* 29, no. 6 (2008): 479–88.

Robins, Lee N., Darlene H. Davis, and David N. Nurco. "How Permanent Was Vietnam Drug Addiction." *American Journal of Public Health* 64, no. 12 Suppl (1974): 38–43.

Rogers, Adam. *Proof: The Science of Booze.* Mariner Books, 2015.

Rorabaugh, W. J. *The Alcoholic Republic: An American Tradition.* Oxford University Press, 1981.

Rosenberg, R. S. "Abnormal Is the New Normal: Why Will Half of the US Population Have a Diagnosable Mental Disorder?" *Slate* (2013).

Ross, Melissa M., Amelia M. Arria, Jessica P. Brown, C. Daniel Mullins, Jason Schiffman, and Linda Simoni-Wastila. "College Students' Perceived Benefit-to-Risk Tradeoffs for Nonmedical Use of Prescription Stimulants: Implications for Intervention Designs." *Addictive Behaviors* 79 (2018): 45–51.

Roth, Tania L., Farah D. Lubin, Monsheel Sodhi, and Joel E. Kleinman. "Epigenetic Mechanisms in Schizophrenia." *Biochimica et Biophysica Acta (BBA)—General Subjects* 1790, no. 9 (2009): 869–77.

Rowe, David C., and Bill L. Gulley. "Sibling Effects on Substance Use and Delinquency." *Criminology* 30, no. 2 (1992): 217–34.

Rubin, Charles. *Don't Let Your Kids Kill You: A Guide for Parents of Drug and Alcohol Addicted Children.* New Century Publishers, 2007.

Rush, Benjamin. *An Inquiry Into the Effects of Spirituous Liquors on the Human Body: To Which Is Added, a Moral and Physical Thermometer.* Thomas & Andrews, 1983.

Rutten, Bart P. F., Eric Vermetten, Christiaan H. Vinkers, Gianluca Ursini, Nikolaos P. Daskalakis, Ehsan Pishva, Laurence de Nijs, Lotte C. Houtepen, Lars Eijssen, and Andrew E. Jaffe. "Longitudinal Analyses of the DNA Methylome in Deployed Military Servicemen Identify Susceptibility Loci for Post-Traumatic Stress Disorder." *Molecular Psychiatry* 23, no. 5 (2018): 1145.

Sargent, James D., Thomas A. Wills, Mike Stoolmiller, Jennifer Gibson, and Frederick X. Gibbons. "Alcohol Use in Motion Pictures and Its Relation with Early-Onset Teen Drinking." *Journal of Studies on Alcohol* 67, no. 1 (2006): 54–65.

Sartor, Carolyn E., Alison E. Hipwell, and Tammy Chung. "Alcohol or Marijuana First? Correlates and Associations with Frequency of Use at Age 17 Among Black and White Girls." *Journal of Studies on Alcohol and Drugs* 80, no. 1 (2019): 120–28.

Sartor, C. E., A. Agrawal, M. T. Lynskey, A. E. Duncan, D. Grant, E. C.

Nelson, P. A. F. Madden, A. C. Heath, and K. K. Bucholz. "Cannabis or Alcohol First? Differences by Ethnicity and in Risk for Rapid Progression to Cannabis-Related Problems in Women." *Psychological Medicine* 43, no. 4 (2013): 813–23.

Schinke, S. P., L. Fang, and K. C. A. Cole. "Substance Use Among Early Adolescent Girls: Risk and Protective Factors." *Journal of Adolescent Health* (2008).

Schuckit, Marc A. "A Brief History of Research on the Genetics of Alcohol and Other Drug Use Disorders." *Journal of Studies on Alcohol and Drugs*, Supplement s17 (2014): 59–67.

Schulenberg, John E., and Jennifer L. Maggs. "A Developmental Perspective on Alcohol Use and Heavy Drinking During Adolescence and the Transition to Young Adulthood." *Journal of Studies on Alcohol*, Supplement 14 (2002): 54–70.

Schwartz, Daniel L., Jessica M. Tsang, and Kristen P. Blair. *The ABCs of How We Learn: 26 Scientifically Proven Approaches, How They Work, and When to Use Them*. W. W. Norton & Company, 2016.

Sederer, Lloyd I. *The Family Guide to Mental Health Care*. W. W. Norton & Company, 2015.

Seligman, Martin E. P. *Flourish*. Simon & Schuster, 2012.

———. *Learned Optimism: How to Change Your Mind and Your Life*. Vintage, 2006.

———. *The Optimistic Child: A Proven Program to Safeguard Children Against Depression and Build Lifelong Resilience*. Mariner Books, 2007.

Seppala, Emma. *The Happiness Track: How to Apply the Science of Happiness to Accelerate Your Success*. HarperOne, 2017.

Sharma, Aditi, and Jonathan D. Morrow. "Neurobiology of Adolescent Substance Use Disorders." *Child and Adolescent Psychiatric Clinics* 25, no. 3 (2016): 367–75.

Sharot, Tali. *The Optimism Bias: A Tour of the Irrationally Positive Brain*. Vintage, 2012.

Shattuck, Gary G. *Green Mountain Opium Eaters*. Arcadia Publishing, 2017.

Sheff, David. *Clean: Overcoming Addiction and Ending America's Greatest Tragedy*. Eamon Dolan/Mariner Books, 2014.

Sheff, David, and Nic Sheff. *High*. HMH Books for Young Readers, 2019.

Sheff, Nic. *We All Fall Down: Living with Addiction*. Little, Brown Books for Young Readers, 2012.

Sher, Kenneth J., Bruce D. Bartholow, and Shivani Nanda. "Short- and

Long-Term Effects of Fraternity and Sorority Membership on Heavy Drinking: A Social Norms Perspective." *Psychology of Addictive Behaviors* 15, no. 1 (2001): 42.

Sher, Kenneth J., Bruce D. Bartholow, and Mark D. Wood. "Personality and Substance Use Disorders: A Prospective Study." *Journal of Consulting and Clinical Psychology* 68, no. 5 (2000): 818.

Shonin, Edo, and William Van Gordon. "The Mechanisms of Mindfulness in the Treatment of Mental Illness and Addiction." *International Journal of Mental Health and Addiction* 14, no. 5 (2016): 844–49.

Shriver, Timothy. *Fully Alive: Discovering What Matters Most.* Farrar, Straus and Giroux, 2015.

Siegel, Daniel J. *Aware: The Science and Practice of Presence—The Groundbreaking Meditation Practice.* TarcherPerigee, 2018.

———. *Brainstorm: The Power and Purpose of the Teenage Brain.* TarcherPerigee, 2015.

———. *The Developing Mind, Second Edition: How Relationships and the Brain Interact to Shape Who We Are.* Guilford Press, 2015.

———. *Mindsight: The New Science of Personal Transformation.* Bantam, 2010.

Siegel, Daniel J., and Tina Payne Bryson. *The Whole-Brain Child: 12 Revolutionary Strategies to Nurture Your Child's Developing Mind.* Bantam, 2012.

Siegel, Daniel J., and Mary Hartzell. *Parenting from the Inside Out: How a Deeper Self-Understanding Can Help You Raise Children Who Thrive: 10th Anniversary Edition.* TarcherPerigee, 2013.

Sinha, Rajita. "Chronic Stress, Drug Use, and Vulnerability to Addiction." *Annals of the New York Academy of Sciences* 1141 (2008): 105.

Sivertsen, Børge, Jens Christoffer Skogen, Reidar Jakobsen, and Mari Hysing. "Sleep and Use of Alcohol and Drug in Adolescence. A Large Population-Based Study of Norwegian Adolescents Aged 16 to 19 Years." *Drug and Alcohol Dependence* 149 (2015): 180–86.

Sloboda, Zili, Hanno Petras, Elizabeth Robertson, and Ralph Hingson. *Prevention of Substance Use.* Springer, 2019.

Smaldone, Arlene, Judy C. Honig, and Mary W. Byrne. "Sleepless in America: Inadequate Sleep and Relationships to Health and Well-Being of Our Nation's Children." *Pediatrics* 119, no. 1 (2007): S29–37.

Smetana, Judith G. *Adolescents, Families, and Social Development: How Teens Construct Their Worlds.* Wiley-Blackwell, 2010.

————. *Changing Boundaries of Parental Authority During Adolescence.* Jossey-Bass, 2005.

Smith, Gregg. *Beer in America: The Early Years, 1587–1840: Beer's Role in the Settling of America and the Birth of a Nation.* Siris Books, 1998.

Smith, G. T., and M. A. Cyders. "Integrating Affect and Impulsivity: The Role of Positive and Negative Urgency in Substance Use Risk." *Drug and Alcohol Dependence* (2016).

Smith, G. T., and L. Guller. "A Comparison of Two Models of Urgency: Urgency Predicts Both Rash Action and Depression in Youth." *Clinical Psychological Science* (2013).

Smith, Joshua P., and Carrie L. Randall. "Anxiety and Alcohol Use Disorders: Comorbidity and Treatment Considerations." *Alcohol Research: Current Reviews* (2012).

Snyder, C. R., Diane McDermott, William Cook, and Michael A. Rapoff. *Hope for the Journey: Helping Children Through Good Times and Bad.* Percheron Press/Eliot Werner Publications, 2002.

Snyder, C. Richard. *Handbook of Hope: Theory, Measures, and Applications.* Academic Press, 2000.

Snyder, C. R. *Psychology of Hope: You Can Get Here from There.* Free Press, 2003.

Spear, Linda Patia. "The Adolescent Brain and the College Drinker: Biological Basis of Propensity to Use and Misuse Alcohol." *Journal of Studies on Alcohol,* Supplement 14 (2002): 71–81.

Sperry, Rod Meade, and editors of the *Shambhala Sun. A Beginner's Guide to Meditation: Practical Advice and Inspiration from Contemporary Buddhist Teachers.* Shambhala, 2014.

Spiegelman, Erica. *Rewired: A Bold New Approach to Addiction and Recovery.* Hatherleigh Press, 2015.

Stefano, George B., Nastazja Pilonis, Radek Ptacek, and Richard M. Kream. "Reciprocal Evolution of Opiate Science from Medical and Cultural Perspectives." *Medical Science Monitor* 23 (2017): 2890.

Steinberg, Laurence. *Age of Opportunity: Lessons from the New Science of Adolescence.* Mariner Books, 2015.

————. *You and Your Adolescent, New and Revised Edition: The Essential Guide for Ages 10–25.* Simon & Schuster, 2011.

Steinberg, Neil, and Sara Bader. *Out of the Wreck I Rise: A Literary Companion to Recovery.* University of Chicago Press, 2016.

Sterling, S. C. *Teenage Degenerate*. No Bueno Publishing, 2016.

Stoddard, Damon. *Pain Drives Change*. Damon Stoddard, 2016.

Storm, Jennifer. *Blackout Girl*. Simon & Schuster, 2009.

———. *Leave the Light On*. Central Recovery Press, 2010.

Swenson, Sandra. *The Joey Song*. Central Recovery Press, 2014.

Szalavitz, Maia. *Help at Any Cost: How the Troubled-Teen Industry Cons Parents and Hurts Kids*. Riverhead Books, 2006.

———. *Unbroken Brain: A Revolutionary New Way of Understanding Addiction*. Picador, 2017.

Tarter, Ralph E., Levent Kirisci, Ada Mezzich, Jack R. Cornelius, Kathleen Pajer, Michael Vanyukov, William Gardner, Timothy Blackson, and Duncan Clark. "Neurobehavioral Disinhibition in Childhood Predicts Early Age at Onset of Substance Use Disorder." *American Journal of Psychiatry* 160, no. 6 (2003): 1078–85.

Temple, John. *American Pain: How a Young Felon and His Ring of Doctors Unleashed America's Deadliest Drug Epidemic*. Lyons Press, 2016.

Terracciano, Antonio, Corinna E. Löckenhoff, Rosa M. Crum, O. Joseph Bienvenu, and Paul T. Costa. "Five-Factor Model Personality Profiles of Drug Users." *BMC Psychiatry* 8, no. 1 (2008): 22.

Teter, Christian J., Sean Esteban McCabe, Kristy LaGrange, James A. Cranford, and Carol J. Boyd. "Illicit Use of Specific Prescription Stimulants Among College Students: Prevalence, Motives, and Routes of Administration." *Pharmacotherapy* 26, no. 10 (2006): 1501–10.

Thayne, Tim. *Not by Chance: How Parents Boost Their Teen's Success in and after Treatment*. Advantage Media Group, 2013.

Thomas, Huw. "A Community Survey of Adverse Effects of Cannabis Use." *Drug and Alcohol Dependence* 42, no. 3 (1996): 201–7.

Thombs, Dennis L., and Cynthia J. Osborn. *Introduction to Addictive Behaviors, Fifth Edition*. Guilford Publications, 2019.

Tiger, Rebecca. *Judging Addicts: Drug Courts and Coercion in the Justice System (Alternative Criminology)*. NYU Press, 2012.

Tokuhama-Espinosa, Tracey. *Mind, Brain, and Education Science: A Comprehensive Guide to the New Brain-Based Teaching*. W. W. Norton & Company, 2010.

Tracy, Sarah W. *Alcoholism in America*. JHU Press, 2009.

Trull, Timothy J., Lindsey K. Freeman, Tayler J. Vebares, Alexandria M. Choate, Ashley C. Helle, and Andrea M. Wycoff. "Borderline Personality Disorder and Substance Use Disorders: An Updated Review."

Borderline Personality Disorder and Emotion Dysregulation 5, no. 1 (2018): 15.

Turrisi, Rob, Kimberly A. Mallett, Michael J. Cleveland, Lindsey Varvil-Weld, Caitlin Abar, Nichole Scaglione, and Brittney Hultgren. "Evaluation of Timing and Dosage of a Parent-Based Intervention to Minimize College Students' Alcohol Consumption." *Journal of Studies on Alcohol and Drugs* 74, no. 1 (2013): 30–40.

Um, M., Z. T. Whitt, R. Revilla, T. Hunton, and M. A. Cyders. "Shared Neural Correlates Underlying Addictive Disorders and Negative Urgency." *Brain Sciences* (2019).

Ungar, Michael. *I Still Love You: Nine Things Troubled Kids Need from Their Parents*. Dundurn, 2014: 209.

———. *Strengths-Based Counseling with At-Risk Youth*. Corwin, 2006.

Urdan, Tim, and Frank Pajares. *Self-Efficacy Beliefs of Adolescents (Adolescence and Education)*. Information Age Publishing, 2006.

Vader, Amanda M., Scott T. Walters, Bahaman Roudsari, and Norma Nguyen. "Where Do College Students Get Health Information? Believability and Use of Health Information Sources." *Health Promotion Practice* 12, no. 5 (2011): 713–22.

Vaillant, George E. *The Natural History of Alcoholism Revisited*. Harvard University Press, 2009.

van der Kolk, Bessel. *The Body Keeps the Score: Brain, Mind, and Body in the Healing of Trauma*. Penguin Books, 2015.

VanVonderen, Jeff. *Tired of Trying to Measure Up*. Bethany House, 2008.

Veliz, Philip, Carol Boyd, and Sean Esteban McCabe. "Adolescent Athletic Participation and Nonmedical Adderall Use: An Exploratory Analysis of a Performance-Enhancing Drug." *Journal of Studies on Alcohol and Drugs* 74, no. 5 (2013): 714–19.

Ventura, Alicia S., and Sarah M. Bagley. "To Improve Substance Use Disorder Prevention, Treatment and Recovery: Engage the Family." *Journal of Addiction Medicine* 11, no. 5 (2017): 339–41.

Vicary, Judith R., and Christine M. Karshin. "College Alcohol Abuse: A Review of the Problems, Issues, and Prevention Approaches." *Journal of Primary Prevention* 22, no. 3 (2002): 299–331.

Vitaro, Frank, Daniel J. Dickson, Mara Brendgen, Brett Laursen, Ginette Dionne, and Michel Boivin. "The Gene–Environmental Architecture of the Development of Adolescent Substance Use." *Psychological Medicine* 48, no. 15 (2018): 2500–7.

Volkow, N. "Substance Use in American Indian Youth Is Worse Than We Thought." Nora's blog, National Institute on Drug Abuse (2014).

Volkow, N. D., and M. Boyle. "Neuroscience of Addiction: Relevance to Prevention and Treatment." *American Journal of Psychiatry* (2018).

Volkow, Nora D., George F. Koob, and A. Thomas McLellan. "Neurobiologic Advances from the Brain Disease Model of Addiction." *New England Journal of Medicine* 374, no. 4 (2016): 363–71.

Wadolowski, Monika, Delyse Hutchinson, Raimondo Bruno, Alexandra Aiken, Jackob M. Najman, Kypros Kypri, Tim Slade, Nyanda McBride, and Richard P. Mattick. "Parents Who Supply Sips of Alcohol in Early Adolescence: A Prospective Study of Risk Factors." *Pediatrics* 137, no. 3 (2016): e20152611.

Walls, Theodore A., Anne M. Fairlie, and Mark D. Wood. "Parents Do Matter: A Longitudinal Two-Part Mixed Model of Early College Alcohol Participation and Intensity." *Journal of Studies on Alcohol and Drugs* 70, no. 6 (2009): 908–18.

Walsh, Froma. *Strengthening Family Resilience.* Guilford Publications, 2015.

Wandzilak, Kristina, and Constance Curry. *The Lost Years.* Jeffers Press, 2006.

Wechsler, Henry. "Binge Drinking on America's College Campuses." Boston: Harvard University (2000).

Wegner, Daniel M. *White Bears and Other Unwanted Thoughts: Suppression, Obsession, and the Psychology of Mental Control.* Guilford Press, 1994.

Westhoff, Ben. *Fentanyl, Inc.* Thorndike Press Large Print, 2020.

Westreich, Laurence M. *A Parent's Guide to Teen Addiction: Professional Advice on Signs, Symptoms, What to Say, and How to Help.* Skyhorse, 2017.

Whitaker, Holly. *Quit Like a Woman.* Dial Press, 2019.

Whitbeck, Les B., and Brian E. Armenta. "Patterns of Substance Use Initiation Among Indigenous Adolescents." *Addictive Behaviors* 45 (2015): 172–79.

White, Helene R. Nicole Jarrett, Elvia Y. Valencia, Rolf Loeber, and Evelyn Wei. "Stages and Sequences of Initiation and Regular Substance Use in a Longitudinal Cohort of Black and White Male Adolescents." *Journal of Studies on Alcohol and Drugs* 68, no. 2 (2007): 173–81.

Whitesell, Nancy Rumbaugh, Janette Beals, Cecelia Big Crow, Christina M. Mitchell, and Douglas K. Novins. "Epidemiology and Etiology of Substance Use Among American Indians and Alaska Natives:

Risk, Protection, and Implications for Prevention." *American Journal of Drug and Alcohol Abuse* 38, no. 5 (2012): 376–82.

Whitman, Glenn, and Ian Kelleher. *Neuroteach*. Rowman & Littlefield Publishers, 2016.

Widom, Cathy Spatz, and Susanne Hiller-Sturmhofel. "Alcohol Abuse as a Risk Factor for and Consequence of Child Abuse." *Alcohol Research & Health* 25, no. 1 (2001): 52.

Wigand, Petra, Maria Blettner, Joachim Saloga, and Heinz Decker. "Prevalence of Wine Intolerance: Results of a Survey from Mainz, Germany." *Deutsches Ärzteblatt International* 109, no. 25 (2012): 437.

Willis, Colin. *Smart Drugs: The Truth About Nootropics: An Introductory Guide to Memory Enhancement, Cognitive Enhancement, and the Full Effects*. CreateSpace Independent Publishing Platform, 2015.

Wills, Thomas A., John Mariani, and Marnie Filer. "The Role of Family and Peer Relationships in Adolescent Substance Use." In *Handbook of Social Support and the Family*, 521–49. Springer, 1996.

Wills, Thomas Ashby, Grace McNamara, Donato Vaccaro, and A. Elizabeth Hirky. "Escalated Substance Use: A Longitudinal Grouping Analysis from Early to Middle Adolescence." *Journal of Abnormal Psychology* 105, no. 2 (1996): 166.

Wolchik, Sharlene A., Irwin N. Sandler, Sarah Jones, Nancy Gonzales, Kathryn Doyle, Emily Winslow, Qing Zhou, and Sanford L. Braver. "The New Beginnings Program for Divorcing and Separating Families: Moving from Efficacy to Effectiveness." *Family Court Review* 47, no. 3 (2009): 416–35.

Wood, A. P., S. Dawe, and M. J. Gullo. "The Role of Personality, Family Influences, and Prosocial Risk-Taking Behavior on Substance Use in Early Adolescence." *Journal of Adolescence* (2013).

World Health Organization. "European Health Report 2015. Targets and Beyond-Reaching New Frontiers in Evidence."

Yamaguchi, Kazuo, and Denise B. Kandel. "Patterns of Drug Use from Adolescence to Young Adulthood: Ii. Sequences of Progression." *American Journal of Public Health* 74, no. 7 (1984): 668–72.

———. "Patterns of Drug Use from Adolescence to Young Adulthood: Iii. Predictors of Progression." *American Journal of Public Health* 74, no. 7 (1984): 673–81.

Yoast, Richard A., Missy Fleming, and George I. Balch. "Reactions to a

Concept for Physician Intervention in Adolescent Alcohol Use." *Journal of Adolescent Health* 41, no. 1 (2007): 35–41.

Yogi, M. C. *Spiritual Graffiti: Finding My True Path.* HarperOne, 2017.

Zapolski, T. C. B., M. A. Cyders, and G. T. Smith. "Positive Urgency Predicts Illegal Drug Use and Risky Sexual Behavior." *Psychology of Addictive Behaviors* (2009).

Zarse, Jane. *Sober and Pissed Off.* Independently published, 2018.

Zerhouni, Oulmann, Laurent Bègue, and Kerry S. O'Brien. "How Alcohol Advertising and Sponsorship Works: Effects Through Indirect Measures." *Drug and Alcohol Review* 38, no. 4 (2019): 391–98.

Zucker, Robert A., Stephen B. Kincaid, Hiram E. Fitzgerald, and C. Raymond Bingham. "Alcohol Schema Acquisition in Preschoolers: Differences Between Children of Alcoholics and Children of Nonalcoholics." *Alcoholism: Clinical and Experimental Research* 19, no. 4 (1995): 1011–17.

Zucker, Robert A., and Sandra A. Brown. *The Oxford Handbook of Adolescent Substance Abuse.* Oxford University Press, 2019.

Zuckerman, Marvin, and D. Michael Kuhlman. "Personality and Risk-taking: Common Bisocial Factors." *Journal of Personality* 68, no. 6 (2000): 999–1029.

INDEX

Siegel, Daniel J., 109–110, 112, 113–114

single parents. *See* divorce and separation

Skaggs, Tyler, 173

ski shot tables, at colleges, 252–253

sleep
  adolescent need for, 116–117
  communicating with young children about, 192
  drugs and alcohol and interruption of, 13, 57, 116–117, 198
  tips for helping adolescents get enough, 117–118

Smith, Gregg, 30

social acceptance. *See* peer pressure

social media, monitoring of children's use of, 123–124

social-emotional learning programs (SEL), in schools, 212, 214, 226

Solo cups, alcohol consumption and, 239

Spiritual Graffiti (Yogi), 112

sports, substance use and, 185, 240, 247

Steinberg, Laurence, 49, 188–190

stimulants
  adolescent brain and, 65, 67–68
  gender differences and, 145–146

stress
  reframing as protective measure, 114–116
  as risk factor for adolescent addiction, 90
  sedatives and damage to adolescent brains, 56–57
  toxic stress, 82, 85–86

"Stress in America" report, 85

styles, of parenting
  categories of, 125, 126–128
  expectations of and consequences for children, 128–133

Substance Abuse and Mental Health Services Administration (SAMHSA), 105

substance use disorder, 4–5, 41–42, 121–122, 133
  adolescent brain's susceptibility to, 45, 52
  beer and, 28–33
  bidirectional sleep disorders and, 116–117
  difficulties of studying in adolescents, 58
  in families, 1–2, 18, 95–97, 133–134
  gateway hypothesis and, 25–28
  language and, 41–43
  risks to children, 18–22
  see also risk factors, for addiction

successes and failures, self-efficacy development and, 103

Sullivan, Edith, 61

summer, as risk factor for adolescent addiction, 90–91

Sumrok, Daniel, 80

synapses, 47

synaptogenesis, brain development and, 49–51

Syria, beer and, 28–29

Teen Insights into Drugs, Alcohol, and Nicotine: A National Survey of Adolescent Attitudes Toward Addictive Substances (2019 report), 123

teenagers. *See* adolescent brains; adolescents

ABOUT THE AUTHOR

JESSICA LAHEY writes about education, parenting, and child welfare for the *Washington Post*, the *New York Times*, and the *Atlantic* and is the author of the *New York Times* bestselling book *The Gift of Failure: How the Best Parents Learn to Let Go So Their Children Can Succeed*. She is a member of the Amazon Studios Thought Leader Board and wrote the curriculum for Amazon Kids' *The Stinky and Dirty Show*. She lives in Vermont with her husband and two sons.